Europe

TIM FLANNERY

Europe

A Natural History

with Luigi Boitani

ALLEN LANE
an imprint of
PENGUIN BOOKS

ALLEN LANE

UK | USA | Canada | Ireland | Australia
India | New Zealand | South Africa

Allen Lane is part of the Penguin Random House group of companies
whose addresses can be found at global.penguinrandomhouse.com.

Penguin
Random House
UK

First published in Australia by The Text Publishing Company 2018

Tim Flann... ...l-clapper shell;
Nummulit... ...ear skeleton;
Neandertha... ...main: Nopcsa;
Robert Plot's... ...leton by Peteron/
Wikimedia C... . Oreopithecus
bambolii... ...edo/Creative
Commons. ...ce CC BY-SA
3.0. Konik po... ...eative Commons.

Printed and bound in Great Britain by Clays Ltd, Elcograf S.p.A.

A CIP catalogue record for this book is available from the British Library

ISBN: 978–0–241–35807–8

www.greenpenguin.co.uk

To Colin Groves and Ken Aplin, life-long
colleagues, and heroes of zoology.

Contents

III

ICE AGES

2.6 Million–38,000 Years Ago

IV

HUMAN EUROPE

38,000 Years Ago to the Future

A GEOLOGICAL TIME CHART

Time Divisions	Important Fossil Deposits	Years Ago
Holocene Epoch		
		11,764
Pleistocene Epoch		
		2.6 million
Pliocene Epoch	Dmanisi	
		5.3 million
Miocene Epoch	Crete footprints Hungarian iron mine	
		23 million
Oligocene Epoch		
		34 million
Eocene Epoch	Messel Monte Bolca	
		56 million
Palaeocene Epoch	Hainin	
		66 million
Cretaceous Period	Hateg	

Introduction

Natural histories encompass both the natural and the human worlds. This one seeks to answer three great questions. How was Europe formed? How was its extraordinary history discovered? And why did Europe come to be so important in the world? For those, like me, seeking answers it is fortunate that Europe has a great abundance of bones—layer upon layer of them, buried in rocks and sediments that extend all the way back to the beginning of bony animals. Europeans have also left an exceptionally rich trove of natural-history observations: from the works of Herodotus and Pliny to those of the English naturalists Robert Plot and Gilbert White. Europe is also where the investigation of the deep past began. The first geological map, the first palaeobiological studies and the first reconstructions of dinosaurs were all made in Europe. And over the past few years a revolution in research, driven by powerful new DNA studies, along with astonishing discoveries in palaeontology, has enabled a profound reinterpretation of the continent's past.

This history begins around 100 million years ago, at the moment of Europe's conception—the moment when the first distinctively European organisms evolved. Earth's crust is composed of tectonic plates that move imperceptibly slowly across the globe, and upon which the continents ride. Most continents originated in the splitting of ancient supercontinents. But Europe began as an island archipelago, and its conception involved the geological interactions of three continental 'parents'—Asia, North America and Africa. Together, those continents comprise about two-thirds of the land on Earth, and because Europe has acted as a bridge between these landmasses, it has functioned as the most significant seat of exchange in the history of our planet.*

Europe is a place where evolution proceeds rapidly—a place in the vanguard of global change. But even deep in the age of dinosaurs, Europe had special characteristics that shaped the evolution of its inhabitants. Some of these characteristics continue to exert influence today. In fact, some of Europe's contemporary human dilemmas result from those characteristics.

Defining Europe is a slippery undertaking. Its diversity, evolutionary history and shifting borders make the place almost protean. Yet, paradoxically, Europe is immediately recognisable. With its distinctive human landscapes, once-great forests, Mediterranean coasts and Alpine vistas—we all know Europe when we see it. And the Europeans themselves, with their castles, towns and unmistakable music, are every bit as instantly recognisable. Moreover, it is important to recognise that Europeans share a highly influential dreamtime—in the ancient worlds of Greece and Rome. Even Europeans whose forebears were never part of this classical world claim it as their own, looking to it for knowledge and inspiration.

So what is Europe, and what does it mean to be European? Contemporary Europe is not a continent in any real geographical sense.** Instead it is an appendix—an island-ringed peninsula projecting into the Atlantic from the western end of Eurasia. In

---

*The size, shape and locations of these landmasses have changed over time. Africa had Gondwanan connections around 100 million years ago. North America has moved away from Europe over the past 30 million years. India's three million square kilometres was not part of the Asian landmass until around 50 million years ago. At times higher sea levels have reduced the area of all of Earth's landmasses, while at others rifting has expanded and fragmented the various lands (such as when the Arabian Peninsula separated from Africa).

**In a geological sense it is part of the Eurasian Plate.

a natural history, Europe is best defined by the history of its rocks. So conceptualised, Europe extends from Ireland in the west to the Caucasus in the east, and from Svalbard in the north to Gibraltar and Syria in the south.* So defined, Turkey is part of Europe, but Israel is not: the rocks of Turkey share a common history with the rest of Europe, while Israel's rocks originate in Africa.

I am not European—in a political sense at least. I was born in the antipodes—Europe's opposite—as the Europeans once called Australia. But corporeally I am as European as the Queen of England (who, incidentally, is ethnically German). The history of Europe's wars and monarchs were drilled into me as a child, but I was taught next to nothing about Australia's trees and landscapes. Perhaps this contradiction triggered my curiosity. Whatever the case, my search for Europe began long ago, before I had ever touched European soil.

When I first travelled to Europe as a student in 1983 I was thrilled, certain that I was going to the centre of the world. But as we neared Heathrow, the pilot of the British Airways jet made an announcement I have never forgotten: 'We are now approaching a rather small, foggy island in the North Sea.' In all my life I had never thought of Britain like that. When we landed I was astonished at the gentle quality of the air. Even the scent on the breeze seemed soothing, lacking that distinctive eucalyptus tang I was barely conscious of until it wasn't there. And the sun. Where was the sun? In strength and penetration, it more resembled an austral moon than the great fiery orb that scorched my homeland.

European nature held more surprises. I was astonished at the prodigious size of its wood pigeons and the abundance of deer on the fringes

---

*Even this definition is not clear cut, for large parts of Europe south of the Alps include fragments of Africa and ocean crust that have been incorporated into the European landmass.

of urban England. The vegetation was so gentle and green in that moist and cushioned air that its brilliant hue seemed unreal. It had very few spines or harsh twigs—so unlike the dusty, scratchy scrub at home. After days of peering into misty skies and looking at soft-edged horizons, I began to feel that I was wrapped in cotton wool.

I made that first visit to study in the collections of London's Natural History Museum. Soon after, I became curator of mammals at the Australian Museum in Sydney, where I was expected to develop a global expertise in mammalogy. So, when Redmond O'Hanlon, the natural history editor at the *Times Literary Supplement*, asked me to review a book on the mammals of the United Kingdom, I reluctantly took up the challenge. The work mystified me because it failed to mention the two species—cows and humans—that had long pedigrees in the United Kingdom, and which I had found to be abundant there.

After receiving my review, Redmond invited me to visit him at home in Oxfordshire. I feared that this was his kind way of telling me that my work was not up to scratch. Instead, I was given a warm welcome, and we chatted enthusiastically about natural history. Late at night, after a sumptuous meal accompanied by many glasses of Bordeaux, he conspiratorially ushered me into the garden, where he pointed to a pond. We crept to its edge, Redmond signalling silence. Then he handed me a torch, and, among the waterweeds, I spotted a pale shape.

A newt! My first. As Redmond knew, Australia lacks tailed amphibians. I was as awestruck as was P. G. Wodehouse's wonderful creation in the Jeeves novels, fish-faced Gussie Fink-Nottle, who 'buried himself in the country and devoted himself entirely to the study of newts, keeping the little chaps in a glass tank and observing their habits with a sedulous eye'.[1] Newts are such primitive creatures that watching them was like looking into time itself.

From the moment I saw my first newt, to discovering the origins of the Europeans themselves, my 30-year-long journey of investigation into Europe's natural history has been filled with discovery. Perhaps what has amazed me most, as an inhabitant of the land of the platypus, is that Europe has equally ancient and primitive creatures that, though familiar, are underappreciated. Another discovery that astonished me was the number of globally important ecosystems and species that arose in Europe, but which are now long-gone from that continent. Who would have guessed, for example, that Europe's ancient seas played an important role in the evolution of modern coral reefs? Or that our first upright ancestors evolved in Europe, not Africa? And who imagines that much of Europe's Ice-Age megafauna survives, hidden away like the elves and fairies of folklore, in remote enchanted woods and plains, or as genes perpetually slumbering in the permafrost?

So much that shaped our modern world began in Europe: the Greeks and the Romans, the Enlightenment, the Industrial Revolution, and the empires that by the nineteenth century had divided up the planet between them. And Europe continues to lead the world in so many ways: from the demographic transition to the creation of new forms of politics and the reinvigoration of nature. Who knew that Europe, with its population of almost 750 million people, is home to more wolves than exist in the USA, including Alaska?

And perhaps most astounding, some of the continent's most characteristic species, including its largest wild mammals, are hybrids. To those used to thinking about 'pure-bloods' and 'mongrels', hybrids are often seen as nature's mistakes—threats to genetic purity. But new studies have shown that hybridisation is vital to evolutionary success. From elephants to onions, hybridisation has allowed the sharing of beneficial genes that have enabled organisms to survive in new and challenging environments.

Some hybrids possess a vigour and aptitude not seen in either parent, and some bastard species (as hybrids are sometimes known) have survived long after the extinction of their parent species. The

Europeans themselves are hybrids, created about 38,000 years ago when dark-skinned humans from Africa began interbreeding with pale-skinned, blue-eyed Neanderthals. Almost the instant those first hybrids appear, a dynamic culture emerges in Europe, whose achievements include the creation of the first pictorial art and human figurines, the first musical instruments, and the first domestication of animals. The first Europeans were, it seems, very special bastards indeed. But long before that, European biodiversity would be destroyed and re-made three times over as celestial and tectonic forces shaped the land.

Let us now embark on a journey to discover the nature of this place that has so influenced the world. In order to do so, we will need several European inventions: James Hutton's discovery of deep time, Charles Lyell's founding principles of geology, Charles Darwin's elucidation of the evolutionary process, and H. G. Wells' great imaginative innovation, the time machine. Prepare to step back in time to that moment when Europe developed its first glimmerings of distinction.

I

THE TROPICAL
ARCHIPELAGO

100–34 Million Years Ago

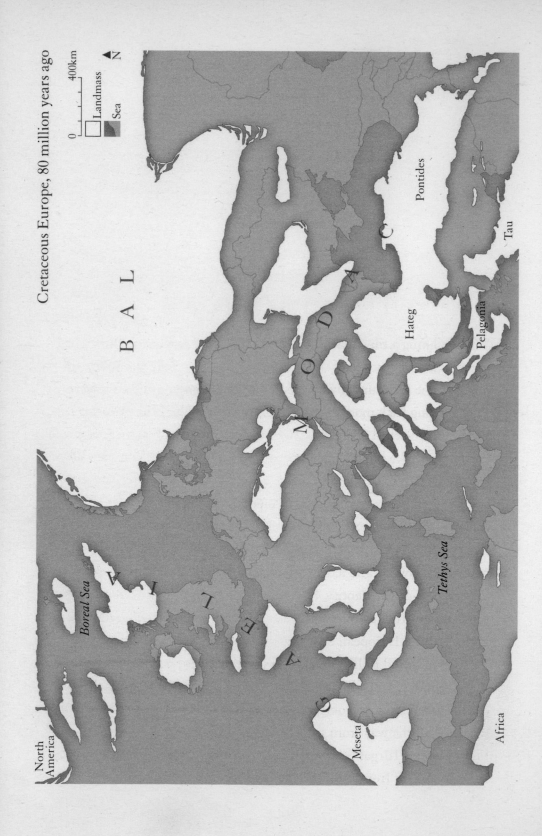

Cretaceous Europe, 80 million years ago

North
America

Boreal Sea

Tethys Sea

Meseta

Africa

Pontides

Hateg

Pelagonia

Tau

B A L

M O D A C

G A E L I A

N

400km

0

Landmass

Sea

CHAPTER 1

Destination Europe

When piloting a time machine you must set two coordinates: time and space. Parts of Europe are unimaginably old, so there are plenty of options. The rocks underlying the Baltic states are some of the oldest on Earth, dating back more than three billion years. Then, life consisted of simple, single-celled organisms, and the atmosphere had no free oxygen. Fast forward 2.5 billion years and we're in a world with complex life, but the surface of the land remains barren. By about 300 million years ago, the land has been colonised by plants and animals, but none of the continents have broken free from the great landmass known as Pangea. Even after Pangea breaks into two, forming the southern supercontinent of Gondwana and its northern counterpart Eurasia, Europe is yet to become a distinct entity. Indeed, it is not until around 100 million years ago, in the last phase of the age of dinosaurs (the Cretaceous Period), that a European zoogeographic region begins to emerge.

One hundred million years ago sea levels were much higher than they are today, and a great seaway, known as the Tethys (which was created when the supercontinents of Eurasia and Gondwana separated) stretched all the way from Europe to Australia. An arm of the Tethys known as the Turgai Strait was an important zoogeographic barrier, cutting Asia off from Europe. The Atlantic Ocean, where it existed at all, was very narrow. Bounding it to the north was a land bridge that connected North America and Greenland with Europe. Known as the De Geer Corridor, this land bridge passed close to the North Pole, the cold and seasonal darkness limiting the species that could cross. Africa bounded the Tethys to the south, and a shallow sea intruded over much of what is today the central Sahara. The geological forces that would in time tear Arabia from Africa's eastern edge and open the Great Rift Valley (thereby widening the African continent), were yet to begin their work.

The European archipelago of 100 million years ago was positioned where Europe is today—east of Greenland, west of Asia, and centred on a region between 30 and 50 degrees latitude north of the Equator. The obvious place to land our time machine would be the island of Bal (which is today part of the Baltic region). By far the largest and oldest island in the European archipelago, Bal must have played a vital role in shaping Europe's primeval fauna and flora. But frustratingly, not a single fossil from the latter part of the age of dinosaurs has been found in the entire landmass, so that all that we know of life on Bal comes from a few scraps of plants and animals that were washed out to sea and preserved in marine sediments that now outcrop in Sweden and southeastern Russia. It would be useless to land our time machine in such a ghastly blank.[1]

It is important to know, however, that ghastly blanks are the norm in palaeontology. To explain their profound influence, I must introduce Signor-Lipps—not some voluble Italian, but two learned professors. Philip Signor and Jere Lipps joined forces in 1982 to propound an important principal in palaeontology: 'since the fossil record of organisms is never complete, neither the first nor last organism in a given taxon will be recorded as a fossil.'[2] Just as the ancients drew a veil of modesty over the critical moment in the story of Europa and the bull so, Signor-Lipps inform us, has geology veiled the moment of Europe's zoogeographic conception, leaving us to set our time machine dial to between 86 and 65 million years ago, when an exceptionally diverse array of fossil deposits preserves evidence of a vigorous infant Europe. The deposits formed on the island chain of Modac, which lies to the south of Bal. Modac was long ago incorporated into a region encompassing parts of almost a dozen eastern European countries—from Macedonia in the west to the Ukraine in the east. In Roman times, this great sweep of land lay within the two sprawling provinces of Moesia and Dacia, from which its name is derived.

At the time of our arrival, large parts of Modac are being pushed above the ocean waves by the first stirrings of the tectonic forces that will, in time, form the European Alps, while others are slipping beneath the sea. Amid this maelstrom of tectonic activity lies the island of Hateg, a place surrounded by submarine volcanoes that intermittently brake the surface, spewing ash over the land. It has endured for millions of years by the time of our visit, allowing a unique flora and fauna to develop. Approximately 80,000 square kilometres in area, about the size of the Caribbean island of Hispaniola, Hateg is isolated, lying 27 degrees north of the Equator and 200–300 kilometres across deep ocean from its nearest neighbour, Bomas (the Bohemian Massif). Today, Hateg is part of Transylvania, in Romania, and the fossils found there are among the most abundant and diverse of the last part of the age of dinosaurs in all of Europe.

Let us open the door of our time machine and step out onto Hateg, land of dragons. We have arrived at the end of a glorious autumn. The sun shines reassuringly, but at this latitude it is rather low in the sky. The air is tropic-warm, and the fine white sand of a bright beach crunches beneath our feet. The vegetation in our vicinity is a mix of low, flowering bushes, but elsewhere groves of palms and ferns are overtopped by ginkgos, their rich-gold autumnal foliage ripe to drop with the first squall of the coming, mild winter.[3] We also see signs, in the form of large, scoured river valleys originating in the distant highlands, that rainfall is highly seasonal.

On a dry mountain ridge, we spy forest giants that resemble the cedars of Lebanon. Belonging to the now-extinct genus *Cunninghamites*, they are indeed a long-vanished kind of cypress. Nearer to hand, a fern-fringed waterhole is resplendent with waterlilies and lined with trees that bear a striking resemblance to the familiar London plane tree (genus *Platanus*). Waterlilies and plane trees are ancient survivors, and Europe has preserved a surprising number of such 'vegetable dinosaurs'.[4]

Our eye is drawn from the land to the azure sea, where the strand is littered with what look at first to be opalescent truck tyres, complete with corrugated treads. They shine with a strange beauty in the tropical sun. Somewhere far out to sea a storm has killed a school of ammonites—nautilus-like creatures whose shells can exceed a metre in diameter—and wave, wind and current have brought their shells to Hateg's shore.

As we walk the glistening sand we detect a stench. Ahead lies a great, barnacled lump stranded by the receding tide. It is a beast, unlike anything living today—a plesiosaur. The four flippers that once powerfully propelled her now lie flat and motionless on the sand. Protruding from the barrel-like body is an inordinately long neck, at the end of which sits a tiny head, still bobbing in the waves.

Three gigantic, vampire-like figures wrapped in leathery cloaks, each as tall as a giraffe, shamble out of the forest. Evil of eye and immensely muscular, the trio surround the carcass, which the largest effortlessly decapitates with its three-metre-long beak. The scavengers circle, and with savage jabs consume the body. Sobered by the spectacle, we step back to the safety of our time machine.

What we have seen hints at what a strange place Hateg is. The vampire-like beasts are a kind of giant pterosaur known as *Hatzegopteryx*. They, rather than some toothy dinosaur, were the island's top predator. Had we ventured inland we might have encountered their usual prey—an array of pygmy dinosaurs. Hateg was a doubly strange place: strange to us because it dates to a time when dinosaurs ruled the Earth, but strange even within the age of dinosaurs because—like the rest of the European archipelago—it is an isolated land with a highly unusual ecology and fauna.

CHAPTER 2

Hateg's First Explorer

The story of how we came to know about Hateg and its creatures is almost as astonishing as the land itself. In 1895, while the Irish novelist Bram Stoker was writing *Dracula*, a real Transylvanian nobleman, Franz Nopcsa von Felső-Szilvás, Baron of Săcel, sat in his castle, obsessing not over blood, but bones. The bones in question had been a gift from his sister Ilona, who had found them while strolling along a riverbank on the Nopcsa family's estate. They were clearly very, very old. Today the Nopcsa family castle at Săcel is a ruin, but in 1895 it was an elegant two-storey mansion with walnut furniture, a large library and a great entertainment hall, whose grand interior can still be glimpsed through its shattered windows. Although modest by grand European standards, the estate provided sufficient income to allow young Nopcsa to pursue his passion for old bones.

Nopcsa would become one of the most extraordinary palaeontologists who ever lived, yet today he is all but forgotten. His intellectual journey began when he left his castle, the gift of bones in hand, and enrolled in a science degree at the University of Vienna. Working largely alone, he soon established that the bones his sister had found were from the skull of a small, primitive kind of duckbilled dinosaur.[1] Fascinated, the count embarked on his life's work—resurrecting the dead of Hateg.

A polymath, loner and eccentric, Nopcsa saw many things more clearly than others, yet he described himself as suffering from 'shattered nerves'. In 1992 Dr Eugene Gaffney, an unexcelled authority on fossil turtles, remarked of Nopcsa that 'in his lucid periods he directed his mind to research on dinosaurs and other fossil reptiles', but that between these moments of brilliance were periods of darkness and eccentricity.[2] Today, perhaps, Nopcsa would be diagnosed as suffering from bipolar disorder. Whatever his malady, it left him without any sense of etiquette. In fact, he all too often displayed 'a colossal talent for rudeness'.[3]

A telling example was recounted by that pioneer of fossil brain research Dr Tilly Edinger, who made a study of Nopcsa in the 1950s. In his first year at university, Nopcsa had published a description of his dinosaur skull—a considerable accomplishment. And when he met the most eminent palaeontologist of his day, Louis Dollo—also an aristocrat, the youthful count crowed, 'Is it not marvellous that I, so young a man, have written such an excellent memoir?'[4] Later, Dollo would offer a backhanded compliment, recalling Nopcsa as a 'comet racing across our palaeontological skies, spreading but a diffuse sort of light'.[5]

At the University of Vienna, Nopcsa seems to have been left largely unsupervised. Isolated from colleagues, his independence extended even to the invention of a glue to repair his fossils. But he did have one colleague, Professor Othenio Abel, who shared an interest in palaeobiology. Abel was a fascist who founded a secret group of eighteen professors who worked to destroy the research careers of 'Communists, Social Democrats and Jews'. He was nearly murdered when a colleague, Professor K. C. Schneider, attempted to shoot him. When the Nazis came to power Abel emigrated to Germany. Visiting Vienna after the Anschluss in 1939, he saw the Nazi flag flying at the university, and proclaimed it the happiest day of his life. Nopcsa had his own way of dealing with Abel. When Nopcsa fell ill he called Abel to his flat, commanding one of Europe's greatest palaeontologists (who was

nonetheless a commoner), to deliver a worn pair of gloves and a coat
to Nopcsa's lover.[6]

As Nopcsa studied his dinosaurs, a second great passion was stir-
ring in his breast. While roaming the Transylvanian countryside he
had met and fallen in love with Count Draŝković. Two years older than
Nopcsa, Draŝković had been an adventurer in Albania, a place which, a
century after Byron's visit, remained exotic, dark and tribal. Influenced
by his lover's stories, Nopcsa made a number of privately funded trips to
Albania, where he lived among the tribes, learned their languages and
traditions, and even got involved in their disputes. A photograph shows
him in his pomp, armed and dressed in the distinctive tribal regalia
of the Shqiptar warrior. If wildly romantic, Nopcsa was also deeply
inquisitive and a meticulous documenter who was soon recognised as
Europe's foremost expert on Albanian history, language and culture.

While travelling in Albania in 1906 Nopcsa met Bajazid Elmaz
Doda, a shepherd who lived high in Albania's Accursed Mountains.
Nopcsa hired Doda as his secretary, confiding to his journal that Doda
was 'the only person since Count Draŝković who has truly loved me'.[7]
His relationship with Doda would last almost 30 years, and in 1923
Nopcsa honoured him by naming a strange fossil turtle after him:
Kallokibotion bajazidi—'beautiful and round Bajazid'.

The turtle's bones had been found alongside those of dinosaurs
on the family estate. At half a metre in length, *Kallokibotion* was a
medium-sized amphibious creature, broadly similar in appearance
to the pond tortoises seen in Europe today. But *Kallokibotion's* bony
anatomy proved that it was very different from any currently living
species, belonging to an ancient, now-extinct group of primitive turtles,
the last representatives of which were the astonishing Meiolaniforms.

The meiolaniids survived in Australia until the first Aborigines
arrived about 45,000 years ago. The last were enormous, land-going
creatures the size of a small car whose tails had become bony cudgels

while their heads bore large, recurved horns like those of cattle. It seems likely that the first Australians saw off almost the last descendants of Bajazid's 'beautiful and round' turtle. But a few had drifted across the sea to the warm, humid, tectonically active islands of Vanuatu. Sequestered in their hermit kingdom the meiolaniids survived—until their lands were in turn discovered—this time by the ancestors of the Ni-Vanuatu, the people who inhabit Vanuatu today. A dense layer of butchered and cooked meiolaniid turtle bones dating to about 3000 years ago marks the arrival of humans. And so was lost a last trace of the lands of Modac—almost the last echo indeed, of that vanished archipelago.

Bajazid, Albania and fossils were the great constants in Nopcsa's life, and of these three he would fall out of love with just one. His involvement with Albania reached its climax just before the outbreak of World War I, when he hatched an audacious, ill-fated plan to invade the country and become its first monarch.* Despite the distraction, Nopcsa

---

* Albania had been incrementally freeing itself from the dying Ottoman Empire, and in 1913 Europe's great powers held a congress in Trieste to decide who should be appointed king. Nopcsa wrote to the chief of the general staff of the Austro-Hungarian army in Trieste, requesting 500 soldiers clad in civilian garments, along with artillery. He would purchase two small, fast steamships and invade Albania, establishing a regime friendly to the Austro-Hungarian Empire. The campaign, Nopcsa told the general, would be swift and would culminate in a triumphant parade through the streets of the capital Tiranë, led by Nopcsa on a white horse. Not all of Nopcsa's motives appear to have been honourable, as he confided in his diary: 'Once a reigning European monarch, I would have no difficulty coming up with the further funds needed by marrying a wealthy American heiress aspiring to royalty, a step which under other circumstances I would have been loath to take.' The British Foreign Office did not see eye to eye with Nopcsa on the issue, and at their behest the congress chose the German Prince William of Wied to be Albania's first king. When World War I broke out and Albania refused to send troops to support the Austro-Hungarians, King Willie's funds were cut off and he was forced to flee. Albania remained kingless until 1928, when the indigenous King Zog I ascended the throne. The bitterly disappointed Nopcsa wrote to Smith Woodward, his palaeontological colleague at the British Museum (now the Natural History Museum), saying, 'My Albania is dead.'

remained elbow-deep in his palaeontology and in 1914 he produced a work on the lifestyle of the Transylvanian dinosaurs that revolutionised understandings of early Europe.[8] What sets his science apart is that he analysed his fossils as the remains of living creatures that existed in specific habitats and responded to environmental constraints. Nopcsa was, in fact, the world's first palaeobiologist.

Nopcsa demonstrated that Hateg was inhabited by just ten species of large creatures. These included a small carnivorous dinosaur known from two teeth (both subsequently lost), which Nopcsa named *Megalosaurus hungaricus*. *Megalosaurus* is a kind of carnivorous dinosaur whose fossils are indeed common elsewhere in Europe, but in older rocks. Its presence on Hateg looked anomalous, and *Megalosaurus hungaricus* was soon demonstrated to be a rare error by the young scientist.

It is a strange scientific fact, worthy of a small diversion here, that the earliest scientific name of *Megalosaurus* is *Scrotum*. The story starts with the first dinosaur fossil to be described and drawn—by the Reverend Robert Plot, in 1677.[9] His *The Natural History of Oxfordshire* was arguably the first modern natural history in English, and in the fashion of the time it covered everything from Oxfordshire's plants, animals and rocks to its notable buildings and even famous sermons given in its churches. Plot correctly identified the fossil in question as the end of a femur. Perhaps, he mused, it was from an elephant brought to Britain during the supposed visit of Emperor Claudius to Gloucester when (according to Plot) he rebuilt the city 'in memory of the marriage of his fair daughter Gennissa with Arviragus, then king of Britain, where he possibly might have some of his elephants with him'. But vexingly, Plot could find no records of elephants closer to Gloucester than Marseille.*

---

*This intriguing history is, sadly, entirely imaginary.

After a long and learned discourse, Plot concluded that the bone, which was found near a graveyard, may have come from a giant. Like many of his contemporaries, Plot believed that Geoffrey of Monmouth's twelfth century work *The History of the Kings of Britain* was solid fact. And so strong is the pull of the great European dreamtime that Geoffrey of Monmouth began his history with a riff on Virgil, in which Brutus, a descendant of the Trojan Aeneas, arrives on Albion's shore to intermarry with the island's original inhabitants, the 'Giants of Albion', and so founds the British race.

Plot did not give the relic a scientific name, and there matters rested until 1763, when one Richard Brookes reproduced Plot's illustration in his own book, *A New and Accurate System of Natural History*.[10] Brookes, who appears also to have believed Geoffrey of Monmouth,* didn't think that the lump Plot illustrated was part of a bone. Instead, he identified it as pair of prodigious human testicles. With the giants of Albion in mind, and awed, perhaps, at the thought of having discovered the very testicles that begat Britain's first queen, Brookes named the fossil the *Scrotum humanum*. Because he had followed the Linnaean system the name remains scientifically valid. And Brookes's identification was evidently convincing: the French philosopher Jean-Baptiste Robinet claimed that he could discern the musculature of the testes, and even the remnants of a urethra, in the fossilised mass.

By the nineteenth century, belief in the veracity of Geoffrey of Monmouth had waned, and scientific research into dinosaurs had commenced. In 1842 anatomist Sir Richard Owen, a man jealous of the scientific achievements of others and not averse to ignoring earlier names for interesting fossils, coined the term 'Dinosauria'. Whether he knew of the *Scrotum* is not clear, but such was the

---

* This is perhaps excusable. Questioning royal pedigrees has always been a risky undertaking.

palaver surrounding Owen's 'discovery', that Brookes's description was lost for more than a century. Even the bone itself vanished. But Plot's drawing allowed it to be identified with certainty as being from the carnivorous dinosaur *Megalosaurus*, whose remains are not uncommon in Jurassic sediments in Britain.

The science of taxonomy builds on its own history, and, in terms of valid scientific names, the loss of an actual specimen counts for naught. At the heart of the science is a small green book known as *The International Code of Zoological Nomenclature*.[11] Like the laws of succession, taxonomy is governed by a rule of primogeniture, which states that the first legitimately coined scientific name takes precedence over all others.* Regrettably for those who do not like the idea of calling dinosaurs scrotums, the code does not forbid the use of names of body parts. Indeed, the great Linnaeus himself named a tropical flower *Clitorea* for the shape of its bright blue pea flowers. A clause in the bylaws of the code, however, states that if a name has not been used since 1899, it can be considered a *nomen oblitum*, or a forgotten name, and so discarded. Such a designation, however, is discretionary.**

When, in 1970, palaeontologist Lambert Beverley Halstead pointed out that *Scrotum* is a scientifically valid name and the first ever proposed for a dinosaur, a shudder went through the normally stolid taxonomic community. Things may not have been helped by the fact that Halstead seems to have been obsessed with dinosaur sex.

---

* While the code may rule that *Megalosaurus* must be known as *Scrotum*, it has nothing to say about higher-level classifications such as Dinosauria, which are left to the discretion of researchers.

** In the 1970s, two British colleagues of mine seriously considered publishing a scientific paper resurrecting *Scrotum*, and renaming the Dinosauria the Scrotalia. The fact that their names were Professors Bill Ball and Barry Cox had, I presume, nothing to do with their interest in the subject.

His most memorable work is an illustrated compendium of dinosaurian copulatory positions—a sort of reptilian *Kama Sutra*—that includes a 'leg over' manoeuvre by the sauropods, the largest dinosaurs of all—that many consider highly dubious. On at least two occasions Halstead took to the stage, where, with his wife, he demonstrated some of the more arcane postures.[*]

At the end of World War I, Transylvania was ceded by the Austro-Hungarian empire to Romania, and Baron Nopcsa lost his castle, his estates and his wealth. As compensation he was offered the position of director at the magnificent Geological Institute in Bucharest. But the loss was too much, and he spent most of his time lobbying the government to reinstate his fiefdom. In 1919 it acceded, but when Nopcsa returned to Săcel his ex-serfs beat him severely, forcing him to relinquish, for a second time, his patrimony.

Nopcsa was for a time confined to a wheelchair and, feeling his powers slipping away, he had himself 'steinacherised'. The operation, which involved an extreme form of unilateral vasectomy, had been developed by Dr Eugene Steinach as a cure for fatigue and waning male potency.[**] While Nopcsa revelled in its marvellous effect on his sexual performance, it did not rejuvenate the rest of the body, as was evident at the 1928 meeting of the German palaeontological society, when Nopcsa delivered a 'brilliant address' on the thyroid gland of various extinct creatures. Tilly Edinger, who attended the meeting, recalled: 'He was pushed among us, reclining in a wheelchair paralysed from head to toe…ending with the words: "with a weak hand

---

[*] Science journalist Robyn Williams, who was in the audience for one performance, noted that Halstead apparently needed fortification, ordering a pint of gin and tonic at the bar.

[**] Steinach was famous for transplanting the testes of male guinea pigs into females, which induced the females to sexually mount other guinea pigs. He was nominated for a Nobel Prize six times.

I have today tried to pull a heavy curtain, to show you a new dawn. Pull strongly, particularly you younger ones; you will see the morning light increase, and you will witness a new sunrise.'"[12]

Unable to reform his institute, Nopcsa resigned as director and became ever more impoverished. He sold his fossil collection to the British Museum and began travelling Europe on his motorcycle, with Bajazid riding pillion. The end came while Nopcsa was studying earthquakes, and he and Bajazid were living in a flat at Singerstrasse 12 in Vienna. As the great dinosaur expert Edwin H. Colbert described it:

> On the 25th of April 1933, something cracked inside Nopcsa. He gave his friend Bajazid a cup of tea heavily laced with a sleeping powder. He then murdered the sleeping Bajazid, shooting him in the head with a pistol.[13]

Nopcsa wrote a note, then shot himself, so putting an end to his noble lineage. His note explained that he was suffering from 'a complete breakdown in my nervous system'. Idiosyncratic to the end, he left the police instructions that 'Hungarian academics' should be strictly forbidden from mourning him. Dressed in his motorcycling leathers, his cremation was fit for a Viking chieftain.[14] Bajazid, in contrast, was buried in the Muslim section of the local cemetery.

CHAPTER 3

Dwarfish, Degenerate Dinosaurs

Among the bones Nopcsa collected on his family estate were the remains of a sauropod—a ponderous, long-necked dinosaur of the brontosaurus type. It was, however, diminutive compared with its relatives, being only the size of a horse. Among the most abundant species were the small armoured dinosaur (*Struthiosaurus*) and the runty duckbilled dinosaur *Telmatosaurus*, which was just five metres long and weighed 500 kilograms. Hateg Island was also home to a now-extinct three-metre-long crocodile, and of course beautiful, round Bajazid's turtle.

Nopcsa's dinosaurs were not only runty, but primitive as well. In describing them he used terms like 'impoverished' and 'degenerated'.[1] In the early twentieth century, such language was unusual. Other European scientists were claiming that the fossils from their country were the biggest, best or oldest of their kind (sometimes, as in the case of Britain's Piltdown Man, fraudulently so). For example, just before the start of World War I, a gigantic sauropod skeleton was discovered in Germany's East African colonies. It was mounted in Berlin's natural history museum where, as late as the 1960s, old Klaus Zimmerman, a zoologist at the museum, would delight at taking visiting Americans to see it 'to show zem zat zey haven't got a bigger one'.[2]

Indeed, it was not unknown during the age of empires to belittle a foreign nation by suggesting that its creatures were small or primitive. When Comte de Buffon, the father of modern natural history, met Thomas Jefferson in Paris in 1781 Buffon claimed that America's deer and other beasts were undersized, miserable and degenerate, as were America's human inhabitants, of whom he wrote: 'the organs of generation are small and feeble. He has no hair, no beard, no ardour for the female.'[3] Jefferson was incensed. More determined than ever to prove the superiority of all things American, he sent to Vermont for a moose skin and pair of antlers of the largest size and was mortified when only a rancid body was delivered, the hair falling from the skin, along with a pair of antlers of a small individual.

Nopcsa seems to have been devoid of such spurious nationalism. He worked carefully on his specimens, trying to understand why they were smaller than dinosaurs found elsewhere, and was the first scientist to cut thin sections from fossilised bones, revealing that the Transylvanian dinosaurs had grown very slowly. The science of zoogeography was in its infancy, but it was known that islands could act as refuges for superannuated, slow-growing creatures, and that limited resources meant that island creatures could become smaller over the generations. So it was that Nopcsa realised the distinctive features of his fossils could be explained by a single fact: they were the remains of creatures that had lived on an island. He would go on to analyse all of Europe's dinosaur faunas, finding hallmarks of 'impoverishment and degeneracy' across the entire region. On that basis, he argued that all of Europe had been, at the time of the dinosaurs, an island archipelago. This profound insight is the foundation stone upon which all studies of European fossils from the end of the age of dinosaurs is built. And yet Nopcsa was ignored. His lack of European chauvinism, his open homosexuality and erratic personality doubtless added to his problems in finding acceptance.

Not all of Europe's dinosaurs are pygmies. Those that lived during the Jurassic Period (earlier than Nopcsa's dinosaurs) could grow to be very large indeed. But they inhabited a Europe that was part of a supercontinent. Dinosaurs that reached the European islands by swimming across the sea could also be large, though their descendants would get smaller over thousands of generations as they adapted to their island homes.

One splendid example of a full-sized European dinosaur is the bipedal herbivore *Iguanodon bernissartensis*. Thirty-eight articulated skeletons of these ponderous creatures, each up to 10 metres long, were found at a depth of 322 metres in a coalmine in Belgium in 1878. The bones, articulated and mounted by Louis Dollo (he to whom Nopcsa gloated about his first publication), were initially displayed in the fifteenth century Chapelle St George in Brussels, an ornate oratory once belonging to the Prince of Nassau. The display was so impressive that when the Germans occupied Belgium during World War I they resumed excavations in the coalmine and were about to uncover the bone-bearing layer when the Allies retook Bernissart. Work stopped, and, while other efforts were made to reach the fossils, in 1921 the mine flooded and all hope was lost.

With the development of new techniques, palaeontologists have been able to learn far more than Nopcsa ever could about life on Hateg. One of the most important developments was the use of fine sieves to recover the bones of tiny creatures, including primitive mammals. Some, such as the Kogaionids, probably laid eggs and hopped like frogs. The bones of strange amphibians known as albanerpetonids, and the ancestors of the midwife toads, which are among the most ancient of European creatures, have been recovered, as have the bones of python-like snakes known as matsoiids, terrestrial crocodiles with serrated teeth, legless lizards, ancestral skink-like creatures, and whiptail lizards. Both matsoiid snakes and serrated-toothed crocodiles survived in Australia until humans arrived on the island continent. This is a familiar occurrence—old Europe surviving into recent times in Australasia.

In 2002 researchers announced the discovery of Hateg's top predator, *Hatzegopteryx*—the creatures we saw when we stepped out of our time machine.[4] Unlike dinosaurs, *Hatzegopteryx* had responded to island life by becoming a giant, making it perhaps the largest pterodactyl that ever lived. The creature is only known from part of a skull, the upper bone of the wing (the humerus) and neck vertebra, but that is enough to allow palaeontologists to estimate its wingspan at 10 metres and its skull at more than three metres long. *Hatzegopteryx* was large enough to kill Hateg's dinosaurs, and its massive dagger-like beak suggests that it caught its prey much like a stork does.[5] While it may have been capable of flight, it almost certainly spent its time on Hateg crawling about on its wrists, with its great leathery wings folded over its body like a shroud. A gigantic kind of Nosferatu comes to mind. How Nopcsa—and indeed Bram Stoker—would have loved this bizarre creature!

CHAPTER 4

Islands at the
Crossroads of the World

The age-of-dinosaurs fauna of Hateg Island is the most distinctive known. But Hateg is only part of the story of saurian-age Europe. To piece together the whole picture we must travel more widely. Flying south from Hateg's shore, we cross the great tropical expanse of the Tethys Sea. In its shallow waters, now-extinct clams known as rudists formed extensive beds. And marine snails called actaeonellids abounded, the largest of which, shaped like an artillery shell, would fill your hand. The shells of these predatory snails were exceedingly thick. They flourished on the rudist reefs, and, wherever sediments allowed it, they burrowed. So abundant were they that today in Romania entire hillsides—known as snail hills—are made up of their fossils. Along with ammonites and the great marine reptiles such as plesiosaurs, sea turtles and sharks were abundant in the Tethys.

To the north of the archipelago a very different ocean existed. It shared almost no species with the warm Tethys—its ammonites, for example, were entirely different types. The Boreal Sea was not tropical, nor was its water clear and inviting. It was filled with a kind of golden-brown planktonic algae known as coccolithophores, whose skeletons would form the chalks that underlie parts of Britain, Belgium and France today. Most of the coccolithophore remains that form the chalk have been ground up—they must have been eaten and excreted by some as-yet unidentified predator.[1]

If the coccolithophores that abounded in the Boreal Sea were anything like *Emiliania huxleyi* (Ehux), the most abundant coccolithophore living today, then we can know a great deal about the appearance of the Boreal Sea. Where upwellings or other sources of nutrients allow Ehux to abound, it can proliferate, as blooms, to the point that the ocean turns milky. Ehux also reflects light and concentrates heat in the uppermost layer of the ocean, as well as producing dimethyl sulphide, a compound that helps clouds form. The Boreal Sea is likely to have been a fantastically productive place, its milky surface waters filled with organisms feasting on the plankton, while cloudy skies would have shielded all from overheating and damaging ultraviolet radiation.

It is hard to overstate just how unusual Europe was towards the end of the age of the dinosaurs. It was a geologically complex and dynamic arc of islands, whose individual landmasses were made up of ancient continental fragments, raised segments of oceanic crust and land newly minted by volcanic activity. Even at this early stage, Europe was exerting a disproportionate influence on the rest of the world, part of which came from the thinning crust beneath it. As heat came to the surface it raised the floor of the sea into ridges between the islands. And this shallowing, reinforced by the creation of mid-ocean ridges as the supercontinents split up, forced the world's oceans to overflow, changing the outlines of the continents and all but sinking some of the European islands.[2] The long-term trend, however, favoured the creation of more land in what was to become Europe.

Like Caesar's Gaul, the European Archipelago towards the end of the age of dinosaurs can be divided into three parts. The great, northern land of Bal and its southern neighbour Modac comprised the principal one. To its south lay an extremely diverse and fast-changing region we shall call the Sea Isles, which comprised the remote archipelagos of the Pontides, Pelagonia and Tau. More than 50 million years later, they would become incorporated into the lands that today fringe the eastern Mediterranean.

To the west of these two great divisions lay a third part. Scattered in the longitudes between Greenland and Bal was a complex of landmasses. In the absence of a widely accepted name we shall call this region Gaelia (from the Gaelic isles and Iberia). Composed of the Gaelic isles (Proto-Ireland, Scotland, Cornwall and Wales) and towards the African sector of Gondwana, the Gallo-Iberian islands, (encompassing parts of what is today France, Spain and Portugal), it is a diverse region. Let us drop in on two locations in Gaelia where an abundant fossil record exists.

Our time machine splashes down in a shallow sea near what is today Charentes in western France. We find ourselves at the mouth of a small river, its flow stilled by a dry spell. A skink-like lizard (one of the earliest skinks) scuttles away over the sea wrack lining the strand, and in a pool of still, green water we see a ripple. A small, pig-like snout breaks the surface, before sinking back. It's a pig-nosed turtle: a single species survives today, in the larger rivers of southern New Guinea and Australia's Arnhem Land.

As we scan the Gaelian shore we see large side-necked turtles sunning themselves. These peculiar creatures are named after their habit of pulling their heads into their shells by folding their necks sideways. Today, side-necks are found only in the southern hemisphere, where they inhabit rivers and ponds in Australia, South America and Madagascar. But the European fossils are from a most unusual branch of the family known as bothremydids. They are the only side-necks ever to take to saltwater, and they were almost entirely restricted to Europe. In the forests lining the river, we see primitive, dwarfish dinosaurs similar to those found on Hateg, though a different species. A movement in the vegetation betrays the presence of a rat-sized marsupial, very similar in shape to the smaller opossums of South American forests today. It is the first modern mammal ever to reach Europe.

The remains of an even more intriguing Gaelian creature—a gigantic, flightless bird—were found in the Provence-Alpes-Côte d'Azur region of southern France in 1995. It was named *Gargantuavis philoinos*, 'gigantic wine-loving bird', because its fossilised bones were exposed among vineyards near the village of Fox-Amphoux (a place otherwise best known, perhaps, as the birthplace of the French revolutionary leader Paul Barras).

At the time these creatures lived, the island that was to become southern France was slowly rising above the waves. But to its south the island of Meseta (which comprised most of the Iberian Peninsula) was simultaneously sinking. Spain would of course rise again, in a process that would produce the lofty Pyrenees and the fusion of Iberia with the rest of Europe. But 70 million years ago, near modern-day Asturias in northern Spain, there existed a lagoon, and as the land sank, the sea flooded it at high tide, and the bones of alligators, pterosaurs and dwarf titanosaurs (long-necked sauropod dinosaurs) were buried in the sediment. Fossils from elsewhere on Meseta tell us that salamanders lurked in the forests of that sinking isle.

CHAPTER 5

Origins and Ancient Europeans

What was distinctively European at this primeval time? And what of it survives today? Scientists talk of a European 'core fauna', by which they mean animals whose lineages were present throughout the archipelago during the age of dinosaurs. The ancestors of most of this core fauna—which include amphibians, turtles, crocodiles and dinosaurs—arrived over water from North America, Africa and Asia very early on. It might be intuited that Asia was a dominant influence, but the Turgai Strait (part of the Tethys Sea) acted as a profound barrier, so opportunities for migration from Asia were limited. Occasionally, however, volcanic islands rose in the strait, providing stepping stones, and over millions of years various creatures made successful crossings, either carried on rafts of vegetation, or by swimming, drifting or flying from one volcanic island to the next.

The dinosaurs arriving from Asia proved the hardiest immigrants, though zhelestids (primitive insect-eating mammals that resembled elephant shrews) also somehow made it. Bipedal hadrosaurs, great hulking lambeosaurs, certain rhino-like ceratopsians and relatives of the velociraptors—all large and probably able swimmers—had the greatest success. Perhaps ten thousand drowned for every one that hauled itself ashore on a European isle. Within a million years or so their descendants would be counted among the dwarfish dinosaurs of the European archipelago.

The migration route from Asia to Europe was more of a filter than a highway, with just a lucky few possessing the bulk, strength or good fortune to be able to traverse it. Yet profound mysteries remain. Why, for example, didn't the soft-shelled or pancake turtles and the typical tortoises, both of which existed in Asia and are good water travellers, make the crossing? Multitudes of smaller creatures must have washed out to sea occasionally in a storm or flood. But whatever the reason, there is no evidence that any survived to settle on a European island.

Throughout Europe's existence, Africa has repeatedly embraced its northern neighbour, then retreated behind a salty curtain. Towards the end of the age of dinosaurs, great rivers flowed from Africa to Europe, and African freshwater fish entered Europe en masse. Among them were ancient relatives of the piranhas and those popular aquarium fish the tetras, along with garfish and freshwater coelacanths, known as mawsonids. The coelacanth is a large fish related to the tetrapods whose discovery off the east coast of South Africa in 1938 caused global astonishment: it was thought to have been extinct for 66 million years.

Alongside these fish, the first of the modern frogs entered Europe. Known as neobatrachids, the group includes the bullfrogs and true toads that are found across Europe today. These African migrants found a welcoming home in what is today Hungary, where their remains have been found in bauxite mines. Certain side-necked turtles, the python-like matsoiine snakes with their vestigial limbs, the terrestrial serrated-toothed crocodiles, and various dinosaurs also entered Europe from Africa. One carnivorous dinosaur, *Arcovenator*, even appears to have migrated to Europe, via Africa, all the way from India. By 66 million years ago, however, the land bridge with Africa had receded beneath the waves.

As connections with Africa were lost, migrations from North America via the De Geer Corridor gained pace. The world was far warmer then than it is today, but nonetheless a long migration over the polar regions, which (as always) experienced three months of darkness each year, was required to make the crossing. Among the early immigrants were the whiptail lizards, though the European branch of the family has long since died off. It's also possible that the early marsupial, whose teeth were found at Charentes in France, also used the De Geer Corridor.

Various members of the crocodile lineage and dinosaurs related to the strange trumpeting *Lambeosaurus* arrived via the De Geer Corridor late in the age of dinosaurs, at a time when a warming climate might have made the route more hospitable. Overall, however, the De Geer Corridor was too polar with conditions too extreme for much of North America's fauna. Certainly, the fearsome *Tyrannosaurus* and the triple-horned *Triceratops*, among the best known of America's dinosaurs, never trod its boreal soils. Even for the lucky few immigrant species that reached Europe, complex barriers restricted their movement. The European archipelago was riven with seas, and each island had its own unique characteristics, some being too small, or perhaps too dry or otherwise unsuitable, to sustain populations of some kinds of creatures. True enough, a few species did achieve a pan-European distribution, but many remained restricted to one island, or a cluster of islands.* Europe was a receiver of immigrants at this time, but did it give anything to the world? The answer is no: there is no evidence

---

* Among those so restricted were the now-extinct solemydid turtles and the lake-dwelling palaeobatrachid frogs, which were restricted to Gaelia, as was the giant flightless bird *Gargantuavis* and the flesh-eating dinosaurs known as abelosaurids, a kind of salamander, possibly strange burrowing lizards known as amphisbaenids, and relatives of the glass lizards (which originated in North America). Bajazid's lovely round turtle and the terrifying *Hatzegopteryx* in contrast, were unique to Hateg, while the python-like matsoiine snakes, with their rudimentary limbs were only found on islands in the east and west of the European archipelago, but not in the middle.

of any European group spreading to other landmasses during the later
phases of age of dinosaurs. Europe did, however, act as a highway for
some creatures, with primitive mammals and some dinosaurs using
it to cross from Asia to America, and vice-versa. An explanation for
this asymmetry may lie in a biological tendency formulated by Charles
Darwin, who thought that species from larger landmasses are competitively superior, and therefore successful migration is usually from
larger to smaller landmasses. As Darwin noted when discussing more
recent migrations:

> I suspect that this preponderant migration from north to south is
> due to the greater extent of land in the north, and to the northern
> forms having existed in their own homes in greater numbers, and
> having consequently been advanced through natural selection
> and competition to a higher stage of perfection, or dominating
> power, than the southern forms.[1]

Most of Europe's core fauna is now long extinct, but there are a
few unlikely survivors. The most important are the alytids (the family
that includes the midwife toads) and the typical salamanders and
newts (family Salamandridae). These relics of Europe's dawning are
deserving of special recognition, for they are in effect Europe's living
fossils, as precious as the platypus and lungfish.

In March 2017, I visited Voltaire's estate in Ferney-Voltaire, near
Geneva. The first flowers of spring were showing through on the south-
facing slopes, but the woodland was still wet and winter-cold. I turned
a log and saw under it a brown creature, barely 10 centimetres long, its
only colour at this non-breeding time the slightest hint of an orange
stripe running down its back. It was an Italian crested newt (*Triturus
carnifex*), which within weeks would enter a pond and, if male, sprout
an extravagant dragon-like crest, bright spots and vivid black-and-white
facial markings.

The creature belongs to the family Salamandridae, whose 77 species are distributed across North America, Europe and Asia. This wide distribution has long obscured their point of origin, but a study of mitochondrial DNA from 44 species has revealed that the salamandrids first evolved about 90 million years ago on an island in the European archipelago.[2] Perhaps it was Meseta, where the oldest salamandrid fossils on Earth have been discovered. The study also revealed that the gloriously colourful Italian spectacled salamanders diverged from the rest of the salamandrid family while the dinosaurs still lived. Just after the dinosaur extinction, salamandrids reached North America and gave rise to the North American and Pacific newts. Even later, around 29 million years ago, some salamandrids reached Asia, and they in turn gave rise to the fire-bellied newts, paddle-tailed newts, and other Asian types.[3]

It is humbling indeed to realise that the ancestors of that tiny, fragile creature I had seen lurking in the depths of Redmond O'Hanlon's pond in Oxfordshire are part of a group that colonised the Americas from Europe long before Columbus, and east Asia well before Marco Polo. To me they, rather than some empire-building human coloniser, are the real embodiment of European success.

CHAPTER 6

The Midwife Toad

It is a truth that sounds more like a fairytale that a toad lies at the heart of ancient Europe.* Today, the common midwife toad can be found from the lowlands of southern Belgium to the sandy wastes of Spain, making it the most successful and widely distributed member of Europe's oldest surviving vertebrate family, the Alytidae, a group comprising the midwife toads, the disc-tongued frogs, the firebellies and the painted frogs.** Look a midwife toad in the eyes, and you are looking at a European whose ancestors blinked at the terrible *Hatzegopteryx*, and one that has survived every catastrophe that has rocked the world over the past 100 million years. More venerable, and more distinctly European than any other creatures, the alytids are living fossils that should be considered nature's nobility.

Some alytids are diligent fathers—which has doubtless aided their survival. When midwife toads mate, the male gathers up the eggs and

---

*Strictly speaking, the term toad should be restricted to members of the family Bufonidae, of which the common European and natterjack toads are good examples. But common parlance has seen the name applied to any warty tail-less amphibian.

**It is frustrating that both the Asian newts and Asian toads are known as firebellies. Though both firebellies raise an interesting evolutionary question. Why should European colonisers of Asia develop such spectacularly coloured underparts?

winds strands of them around his legs. He can mate up to three times per season, so some individuals carry three broods in this manner. For up to eight weeks the male carefully tends the eggs, carrying them everywhere he goes, wetting them if they are in danger of drying out, and secreting natural antibiotics from his skin to protect them from infection. When he senses that they are about to hatch, he seeks out a cool, calm pond for the tadpoles to grow in.

There are five species of midwife toad: the widespread nominotypical species, three restricted to Spain and its islands, and one that reached Morocco from Spain in the recent geological past. The Majorcan midwife toad has the distinction of being a Lazarus species, being first described from fossils.* It was widespread on Majorca prior to the arrival of humans, but when mice, rats and other predators reached the island the toads vanished. A few survived, undetected, in the deep valleys of the Serra de Tramuntana, and following their discovery in the 1980s they were reintroduced into various parts of the island, where, with a little protection, they are once again thriving.[1]

Midwife toads played a crucial role in an all-but-forgotten scientific debate in the early twentieth century between English statistician and biologist William Bateson—the man who coined the term 'genetics'—and Professor Richard Semon and his colleagues, who argued for non-genetic inheritance via a Lamarckian form of cellular 'memory'.[2]

Richard Semon was a formidable intellect. Born in Berlin in 1859, he spent much of his youth in the wilds of colonial Australia, collecting biological specimens and living with the Australian Aborigines. Upon his return to Germany he studied how ideas and traits are passed from one individual to another. His book *The*

---

* The term 'Lazarus species' was coined by the palaeontologist David Jablonski to describe a taxon that was thought to have gone extinct during a mass extinction event, but which is found to exist several million years later.

Mneme, published in 1904, was a foundation work on the subject, and its influence was destined to be felt far beyond biology. It commences with the observation:

> The attempt to discover analogies between the various organic phenomena of reproduction is by no means new. It would be strange if philosophers and naturalists had not been struck by the similarity existing between the reproduction in offspring of the shape and other characteristics of parent organisms, and that other kind of reproduction we call memory.

In trying to explain his concept, Semon reminisces:

> We were once standing by the Bay of Naples and saw Capri lying before us; nearby an organ-grinder played on a large barrel-organ; a peculiar smell of oil reached us from a neighbouring 'trattoria'; the sun was beating pitilessly on our backs; and our boots, in which we had been tramping about for hours, pinched us. Many years after, a similar smell of oil ecphorised [brought to mind] most vividly the optic engram [memory] of Capri. The melody of the barrel-organ, the heat of the sun, the discomfort of the boots, are ecphorised neither by the smell of the oil nor by the renewed experience of Capri...This Mnemic property may be regarded from a purely physiological point of view, in as much as it is traced back to the effect of stimuli applied to the irritable organic substance. [3]

This was true, according to Semon, regardless of whether the mneme was a memory or an inherited aspect of a body such as eye colour.

British–German rivalry and the horrors of World War I meant that Semon's book was not translated into English until 1921, too late for its author. A great nationalist, he felt the defeat and shame of surrender so acutely that he wrapped himself in the German flag and shot himself. Today, Semon is not entirely forgotten. A skink discovered living on the island of New Guinea bears his name. *Prasinohaema semoni*'s most distinctive attribute is that its blood is bright green.

After Semon's death, his work was continued at Vienna University by a team including the brilliant young scientist Paul Kammerer, who was a student of music before he turned to biology. His experiments look bizarre by modern standards but were considered the height of scientific elegance at the time. His greatest triumphs involved manipulating the love life of the 'obstetric toad' (the common midwife toad). Toiling over hundreds of the warty creatures, he persuaded them to forgo their predilection for having sex on land.

Aquatic copulation was finally achieved by keeping them 'in a room at high temperature…until they were induced to cool themselves in a water trough…here the male and female found each other…' and, Kammerer reported, mated in the normal anuran* manner (where the female releases the eggs into the water, where they are fertilised), rather than in the manner of midwife toads (where the male helps squeeze the eggs from the female, then wraps them around his hind limbs). This was interpreted as the toad 'remembering' the ancestral way of having sex—a trait, it was claimed, which persisted in subsequent generations. The male descendants of midwife toads that mated in water, Kammerer said, even grew a special black wart on their palms that they used to grasp the wet and slippery female—a feature seen in many frogs and toads, but which has been lost in the midwife toads.

Even after producing such astonishing 'proofs' of Semon's mnemic theory, the amphibians in Kammerer's lab were allowed no rest. In a separate experiment, Dr Hans Spemann forced the bombinator

---

* Anurans are amphibians without tails—the frogs and toads.

(firebelly) toad* to grow eye lenses on the back of its head—a remarkable feat, but one that was surpassed by Gunna-Ekman, who induced green tree frogs (*Hyla arborea*) to grow eye lenses anywhere on the body 'with the possible exception of the ear and nose'. This, it was argued, proved that the frog's skin 'remembered' how to grow eyes—if appropriately stimulated. Meanwhile, Walter Finkler devoted himself to transplanting the heads of male insects onto the bodies of females. The hybrid creatures showed signs of life for several days, but, perhaps not surprisingly, exhibited disturbed sexual behaviour.

By the 1920s Kammerer's work was under severe assault, for it flew in the face of the 'neo-Darwinian orthodoxy', then being championed by William Bateson, who was described, when young, as 'snobbish, racist and intensely patriotic'.[4] Bateson's attacks on Kammerer were, according to Arthur Koestler, vitriolic and obsessive. Bateson suspected fraud from an early stage, and fraud, indeed, was proved in 1926, when it was discovered that the pigmented wart on the palms of one of Kammerer's midwife toads had been tattooed onto the skin. To this day, the perpetrator of the fraud remains unknown, but it may have been an assistant who was a Nazi sympathiser trying to discredit Kammerer, who was a Jew, an ardent pacifist and a socialist. The fraud was held up by Bateson as evidence that Kammerer's life's work was suspect and, with his reputation in tatters, Kammerer took a walk in a forest and—like Semon before him—shot himself.

In 2009, the developmental biologist Alexandre Vargas re-examined Kammerer's findings and claimed that, beyond the tattooed toad palm, they may not have been fraudulent, but could be explained by epigenetics—changes caused by the modification of gene expression, rather than alteration of the gene itself. Other researchers have

---

* 'Bombinator' refers to the bumble bee, whose humming flight noise is said to resemble the croaking of this most unusual toad. Its call, incidentally, is produced with an inward rush of air, as opposed to the outwards push used by most other frogs and toads.

claimed that Kammerer should be credited as the founder of the epigenetic phenomena known as 'parent of origin' effects, whereby genetic imprinting allows the silencing of certain genes. A century after their suicidal despair, both Kammerer and Semon are gaining some recognition.

The midwife toads have a close relative in Europe in the firebelly or 'bombinator' toads (the same creatures that Hans Spemann manipulated to grow eye lenses on the backs of their heads). There are eight species of these small but colourful amphibians, and they are the only real travellers among the alytids.* Tens of millions of years ago, these tiny firebellies managed to cross the entire breadth of the Eurasian landmass, and today five of the eight species inhabit mountains and swamps in China.

The alytids are one of just three ancient frog families in the order Archaeobatrachia—the most primitive frogs and toads surviving today. The other two are New Zealand's frogs, and the tailed frogs of the North America's Rocky Mountains. Combined, these two families contain just five species, while the alytids include about 20 living species, over half of which inhabit Europe. The alytids include six species of disc-tongued frogs, two of which have reached North Africa, and the painted frogs (*Latonia*), of which there is but a single living species. Between 30 million and one million years ago painted frogs abounded in Europe, but then they became extinct. In 1940 biologists collected two adult frogs and two tadpoles in the vicinity of Lake Hula, in what is now Israel. To everyone's astonishment, they were painted frogs. The larger of the two promptly ate its smaller companion, and in 1943 the cannibal—by then pickled in preserving fluid in a university collection, was pronounced a new species, the Hula painted frog.

---

* Their position in the family Alytidae is still debated, with some researchers placing them in their own family, the Bombinatoridae. Nobody doubts, however, that they are close relatives of the alytids.

One more painted frog was collected in 1955, but after that the creatures vanished, and by 1996 the International Union for the Conservation of Nature presumed them to be extinct. Israel, however, continued to list the species as endangered. Its faith was repaid in 2011 when a living painted frog was located by ranger Yoram Malka in the Hula Nature Reserve in northern Israel, where a population of several hundred survives. The Hula painted frog is the ultimate Lazarus taxon: thought extinct for a million years, it was discovered clinging to life in a swamp on Europe's periphery.

Until half a million years ago, the alytids shared Europe with another group of amphibians, the palaeobatrachids.[5] Frogs generally do not make good fossils, but the palaeobatrachids are an exception, and you can see their exquisitely preserved remains on exhibition in many European museums. In habits and body shape, the palaeobatrachids resembled the grotesque clawed frogs of Africa and the Surinam toads of South America, and like them seem to have lived their entire lives underwater, with a preference for lakes, including deep, still ones where the chances of having been preserved as a fossil are far better than for those living in swamps or on land. We missed out on seeing these frogs in the flesh by the merest whisker of geological time.

This Europe of 'in the beginning' may seem like a distant place, with more in common with, say, Australasia than with the Europe of today, but even at this early stage there are some threads that link it with the Europe of more recent times. One is its extremely diverse nature. In the beginning, it was the great lumbering reptiles that differed across the European islands. Today it is distinct languages and human cultures that exist within and across various boundaries. But, just as importantly, then as today, Europe was a land of exceptional dynamism and large-scale immigration—of species that would arrive and find a place among Europe's existing inhabitants, adapt to local conditions and help make Europe anew.

CHAPTER 7

The Great Catastrophe

In a thick sandstone layer in the Tremp Basin in the southern Pyrenees, a ghostly shadow of Europe's last dinosaurs can be seen—in the form of footprints.* Because the rocks preserving them have been lifted, folded and eroded from below, many footprints are preserved in the roofs of overhangs, so that what we see is a great stony replica of the dinosaur's foot, stomping down on us from above.[1] The prints were mostly made by long-necked sauropods and bipedal hadrosaurs which had migrated into the archipelago from North America and Asia towards the end of the age of dinosaurs. Where they were going, and where they came from on that particular day, nobody knows. But we do know that within 300,000 years of the prints being made the descendants of the creatures that made them would be swept off the face of the Earth. Rare evidence of the cataclysm that destroyed them is preserved in the rocks of the Tremp Basin, where a succession of sediments accumulated unbroken over a long period both before and after the extinction event.

The cause of the extinction of the dinosaurs has been long debated. Some palaeontologists argued that climatic or geological changes

---

* In geology, a basin is an area of down-folded or down-faulted rocks that has accumulated a thick layer of sediments.

interrupted the dinosaurs' food supply, but no one could convincingly explain what had happened until, in 1980, a team of researchers—led by the physicist Louis Alvarez and his geologist son Walter—suggested that an asteroid had struck the Earth, causing a nuclear winter severe enough to trigger a mass extinction event. The team announced that they had evidence of this—in the form of sedimentary layers rich in iridium derived from the asteroid—in rocks from around the globe. Building on this pioneering work, in 2013, a team led by Professor Paul Renne of the Berkeley Geochronology Center used argon dating to pinpoint the moment of the strike: 66,038,000 years ago, give or take 11,000 years.[2]

Some palaeontologists seemed outraged by the bolide theory of extinction or—more accurately, perhaps—that a person from outside their discipline dared dabble in their business. They argued that the dinosaurs existed for millennia after the impact, or that at the time of the disaster they were in gentle decline anyway. Others denied that an asteroid impact could have such a catastrophic effect.[3] Despite counter-arguments, it is now accepted that some sort of heavenly body (a bolide) struck Earth and caused the extinctions. Increasingly, scientists are convinced that the offending object was a meteorite or comet of about the size of Manhattan Island.

So how bad could an asteroid strike be? One answer comes from the fact that it takes a great deal of force to shock quartz. Indeed, until recently, it wasn't believed that quartz was shockable. Then scientists examined some sand grains from the vicinity of an underground nuclear test. The power of the blast had been sufficient to deform the quartz's crystal structure, which showed up as microscopic lines in the grains. It takes more than two gigapascals (two billion pascals) of pressure to shock quartz in this manner (for comparison, the atmosphere at sea level exerts a little over 100,000 pascals of pressure). Volcanoes, incidentally, don't shock quartz. Even though they can generate the required pressure, shock requires temperatures to remain relatively low, and volcanoes are too hot. The bolide that exterminated the dinosaurs released two million times more energy than the largest nuclear test ever conducted, creating the greatest volume of shocked quartz in the history of our planet: the stuff is ubiquitous in rocks formed at that time.

The bolide struck near the Equator, on what is now the Yucatan Peninsula in Mexico. The impact displaced about 200,000 cubic kilometres of sediment, and the shockwaves would have rung Earth like a bell, triggering volcanic eruptions and earthquakes globally.[4] The mega-tsunami it caused is estimated to have been several kilometres high—one of the largest in the planet's history, and must have still been substantial by the time it reached the European Archipelago. In the aftermath, flaming debris rained from the sky, triggering fire-storms that consumed forests wholesale, leaving behind great layers of charcoal. Because oxygen levels were higher then, even wet vegetation would have burned.*

When the fires died down, a nuclear winter, initiated by particles blasted into the atmosphere that obscured the sun, commenced. To add to the destructive effect, the bolide had landed in a bed of gypsum, creating enormous volumes of sulphur trioxide that mixed with water to produce sulphuric acid, reducing the amount of sunlight reaching Earth by as much as 20 per cent, exacerbating the nuclear winter so that it caused freezing temperatures and prevented photosynthesis for about a decade. Paradoxically, the nuclear winter was followed by global warming caused by CO_2 released by fires and volcanic activity. Ocean circulation would have ceased abruptly and may have remained severely impaired for thousands of years. Marine life was devastated. Never again would the world see the likes of the glorious ammonite,

---

*Oxygen dropped to near present levels after the impact.

or the ungainly plesiosaur. Nor would it have the rudist clams or the artillery-shell-like actaeonellids.

The impact site was relatively close to the European Archipelago, and we can anticipate that the consequences of tsunami and wild-fire there were severe. Nowhere on Earth could have escaped the nuclear winter that followed. Almost everything weighing more than a few kilograms—including Europe's stunted dinosaurs and Bajazid's turtle—became extinct. Even many smaller creatures, including Europe's whiptail lizards, its matsoiid snakes and some primitive mammals, also vanished. *Sic transit gloria mundi!*

Europe's freshwaters, however, provided important refuges. Its amphibians came through largely unscathed, as did some of its aquatic turtles. Deep water buffers extremes of heat and cold, and freshwater ecosystems can survive for a time without photosynthesis, because bacteria and fungi feeding off detritus washing in from the devastated land provide a base for the food chain. Eventually, at the top of the food chain, frogs and turtles can scavenge. So it was that the ancestors of the delicate salamanders and the midwife toads survived global catastrophe.

Frustratingly, we have almost no European fossils from the moment of the bolide strike to tell us what was happening on land. We are more fortunate when it comes to the seas. In Italy and the Netherlands, among other places, the precise moment of the strike can be seen—and touched—in stone. Indeed, it was at Gubbio in Italy's Apennines that the iridium layer was first identified and studied, after being splendidly exposed in a roadside cutting. The layer proved to be rich in small glassy spheres—the remains of rocks that had been melted and shot out of Earth's atmosphere, before solidifying and raining back down.

Perhaps the most impactful of marine extinctions, at least in Europe, was that of the coccolithophores, whose fossils, deposited by the gigatonne, formed the chalk that gives the Cretaceous Period its name. From the white cliffs of Dover to the chert used for building and to the rocks of the tunnels of World War I battlefields in Belgium and northern France, Europe is filled with evidence of the coccolithophores' past abundance. With the extinction of many key types, the creta (chalk) would never form again.[*]

Although most of us are oblivious to the threat, asteroid strikes continue to be a possibility. In December 2016, NASA scientists warned that we are 'woefully unprepared' in the event of an asteroid or comet striking Earth.[5] Even a far smaller strike than the one of 66 million years ago could devastate our civilisation.

---

[*] Some coccolithophores must have survived, because chalk deposition continued in a few places, including England and Denmark, for a few million years after the bolide impact.

CHAPTER 8

A Post-Apocalyptic World

The great bolide extinction event marks the end of the age of dinosaurs and the beginning of the age of mammals. Known as the Cenozoic Era, meaning 'recent life', it's the division of time that we live in. The Cenozoic is divided into epochs, beginning with the Palaeocene, which extends from about 66 to 56 million years ago.[1] Meaning 'old new', this confusing name was coined in 1874 by Wilhelm Philippe Schimper, a French expert on mosses who also dabbled in palaeobotany.

What was the European Archipelago like once the climate settled down, and life began to reclaim the land? Frustratingly, we face a ghastly blank in the fossil record at this critical moment—a blank that persists for five million years. The chance that fossils of land-based creatures would be preserved was not helped by the fact that much of the European Archipelago was submerged at the time (though large islands did exist). But from evidence elsewhere, particularly in North America, we can assume that for millennia a devastated landscape dominated by ferns existed.* Then, slowly, the surviving trees and bushes emerged from their refuges, perhaps from deep valleys or the seed bank in the soil, or from seeds that drifted across the ocean. But the climate was

---

*Some ferns are pioneer species, able to quickly colonise bare ground.

now altered: Europe was cooler and drier, so new types of plants flourished, while some survivors now found conditions difficult.

Despite the changed climate, how the trees must have grown! For they were not only freed from the browsing lips of the dinosaurs, but from the mouths of many leaf-eating insects that, at least in North America, had become extinct as well.[2] It seems reasonable to assume a similar impact in Europe, which would have allowed the island forests to grow denser and more quickly than ever before. Reproduction, however, may have been more difficult, for pollinators and seed dispersers were in short supply.

What was life like in those rapidly growing forests? We gain our insight courtesy of a hole, 25 metres deep and just a metre wide, dug in a football field at Hainin, near Mons in Belgium, which intersects sediments laid down about five million years after the bolide impact. The excavation resulted from a chance discovery in the 1970s when geologists drilled several smaller holes hoping to obtain samples of marine sediments. Instead they found something infinitely more valuable: fossils of the earliest European land-based creatures from the age of mammals.[3] Subsequently, three other holes were sunk in the football field, each yielding new fossils and new insights into a vanished age.

In a brief moment of glory just before the drilling frenzy, the ROC de Charleroi-Marchienne played in First Division, but today it lingers in Third Division B. I hope that the drilling of the football field had nothing to do with it, but, speaking for myself, I would have dug up half of Brussels itself for the fossils that those drillers got at Hainin. Admittedly, the volume of finds was rather meagre. The 400 fragments—mostly isolated teeth from rat-sized mammals, and a few bones of reptiles, amphibians and fish—would fill a matchbox or two. But what a yield of information! They tell us that the fresh waters at Hainin must have been substantial, for they included the remains of a large predatorial fish known as a bonytongue, or saratoga. Much sought after

by game fishers, today they are only found in the rivers of southeast
Asia and Australia, but at the time the Hainin deposit was forming
they occurred all over the world.[4] The bones of ancient alytids—ancestors of the midwife toads—were also present, as were the remains of
a salamander.

The albanerpetonids—was there ever a more awkward name? Let's
call them the pert'uns—were newt-like amphibians that burrowed
through leaf litter. Their fossils are found in North America, Asia and
Europe (including at Hainin), where they occur in sediments formed
both before and after the bolide strike. Imagine a pert'un lying in the
palm of your hand. An inhabitant of the soil, it is probably dark in
colour, and might be mistaken for a knobbly skinned newt. But unlike
any newt, pert'uns feel hard, because under their skin they have bony
armour. The creature raises its head to look at you, revealing a lithe
and flexible neck unlike that of any living amphibian.

Amphibians were the first vertebrate colonisers of the land—back
in the Devonian Period, some 370 million years ago. Today we have
just three major lineages of amphibians—the anurans (frogs and toads),
the newts and salamanders, and the worm-like caecilians, all of which
can trace their ancestry back to long before the dinosaurs. The pert'uns
were a fourth lineage—one that originated at the very beginning of
the amphibian story. Over the generations, the eyes of pert'uns have
taken in most of the history of life on land. And we humans almost
got to see them. In 2007, fossils dating to a mere 1.8 million years ago
and preserved in deposits formed in limestone near Verona, Italy, were
recognised.[5] To be robbed of the chance of meeting a pert'un by such
a small span of time (geologically speaking), after they'd been around
for 370 million years, seems tragic. It would be lovely to think that in
some obscure valley in Europe, a pert'un survives today.

It seems very odd that the eggshells of two different kinds of turtles
were preserved at Hainin, as eggs rarely fossilise. We cannot identify

the turtles that produced the eggs, but the fact that three of the four great European turtle lineages became extinct when the bolide hit limits the possibilities. The only survivors were the side-necks, but they were on borrowed time, becoming extinct about 10 million years later. All European turtles living today are descended from immigrants that arrived after the bolide impact.

Two different crocodile-like creatures are represented by a vertebra each, so little can be said about them.[6] But two other tiny vertebrae attest to something far more interesting: the presence of a blind snake. Blind snakes are the most primitive of all snakes, and the Hainin bones are the oldest fossils of a blind snake found anywhere on Earth.[7] Burrowing creatures, they live like worms, which they closely resemble, and feed on ants and termites. A single species, found in the Balkans and the Aegean Islands, survives in Europe today.

Fossils of amphisbaenids—bizarre, subterranean, worm-like lizards that originated in North America more than 100 million years ago—were also found at Hainin. About 10 centimetres long, they are formidable predators with horrible-looking eyeless heads and powerful, interlocking teeth that can tear hunks off their victims, which are eaten alive. With loose skin that seems to move along of its own accord and drags the body behind it, an amphisbaenid can move forwards or backwards with equal ease. Blind, pale of skin and uncanny, some bear a resemblance to the Seer of Kattegat in the television series *Vikings*. The amphisbaenids survived the bolide impact in North America, and their presence at Hainin indicates that they migrated to Europe very early on.[8] Unlikely seafarers, they appear to have crossed the north Atlantic on drifting vegetation.[9] Today, four species of amphisbaenids survive in Europe—two on the Iberian Peninsula and two in Turkey.[*]

---

[*] The ancestors of Europe's amphisbaenids appear to have arrived across the sea in separate migrations.

A most striking thing about Hainin's fauna is how chthonic it is. Salamanders and toads, sightless, burrowing lizards and blind snakes are all creatures of the earth itself. As I think about their world, I'm reminded of images of Europe in the wake of a far more recent catastrophe. Film footage from the end of World War II captures poor, beleaguered creatures emerging from their burrows among the rubble into a devastated and sadly reduced world. It is as if only the bowels of the Earth itself could offer refuge from such destruction.

Sixty-six million years ago, the consequences of the bolide impact lasted not decades but for millions of years. Yet life did eventually recover. In a forest beside the sea, as palaeontologists believe the Hainin site once was, those regrowing groves were enlivened by one group of small survivors. Scrambling over fallen logs and up into the branches was a surprising diversity of rat-sized mammals. The most abundant were fifteen-centimetre-long nocturnal eaters of insects and fruit known as adapisoriculids. They were long thought to be related to hedgehogs, but more recent studies identify them as primitive creatures that did not develop a placenta, but which in other respects were similar to the placental mammals. They looked much like rats, and existed for about 10 million years after the bolide strike. Most species were European.

Among the most intriguing mammals lurking in Hainin's forests were the kogaionids—initial survivors of the bolide impact, we met them briefly on Hateg. Unique to Europe, their remains abound at Hainin; one kind *Hainina*, takes its name from the site. Great survivors they may have been, but the kogaionids were very primitive mammals that probably laid eggs. Although not much larger than rats, the kogaionids could never have been mistaken for rodents. Let us imagine that we are in Hainin's ancient forests. A movement in the umbrageous undergrowth betrays the presence of something leaping from the ferns. It moves about just like a frog but is covered with fur.

Behold a kogaionid, the only mammal ever to have developed a means of locomotion akin to that of frogs and toads.[10]* As it opens its mouth to consume the blind snake it has ambushed, you see large shearing premolars, which it uses to cut up its prey. Strangely, the long lower incisors it has impaled its victim on are blood-red—a result of their enamel being strengthened with iron.[11] Rat-sized, primitive ungulates, marsupials and elephant shrews complete the Hainin mammal fauna.[12] All could have survived the bolide impact in burrows and, during the dark chill that followed, by eating small invertebrates such as worms, hoppers and insects, or the seeds left behind in the soil.

---

*There is still some debate about the locomotion of the kogaionids, Earlier reconstructions depicted them to be squirrel-like creatures, but a recent re-analysis suggests that they moved like frogs.

CHAPTER 9

New Dawn, New Invasions

Ten million years after the extinction of the dinosaurs a new geological epoch was dawning. The onset of the Eocene is marked by a shift in the ratio of two isotopes of carbon—C^{12} and C^{13}—indicating an eruption of fossil carbon into the atmosphere. The event is one of the most striking in Earth's history. Within 20,000 years—a mere geological instant—the fossil carbon caused global temperatures to increase by between 5°C and 8°C and they remained elevated for 200,000 years. At the same time the oceans, especially the north Atlantic, acidified. Ocean circulation changed radically (in some regions reversing), and deep-sea foraminifera (single-celled organisms) went extinct *en masse*. On land, rainfall patterns changed, with some regions subjected to biblical deluges and others drying up. Erosion and leaching on an unprecedented scale depleted the soil, laying down vast new sediment beds on river floodplains. Rainforests flourished as far north as Greenland.

Some researchers think that the warming was caused as kimberlite pipes (volcanic vents originating deep in the Earth's mantle) reached the surface near Lac de Gras in northern Canada and released huge amounts of carbon. Others think that a release of natural gas from the ocean depths was the cause. The extreme acidification of the central and north Atlantic supports this idea, as does the presence of several large crater-like structures on the ocean floor, which have narrow sheets of volcanic rock, known as sills, at their bases. The molten rock in the sills may have ignited vast reserves of shallowly buried natural gas, much like a match applied to a gas barbecue.[1] Whatever the cause, it is generally accepted that the warming was triggered by a smaller annual flow of carbon than humanity is currently contributing to the atmosphere.[2]

The name Eocene (meaning 'new dawn') was coined by the father of modern geology, Charles Lyell. His three-volume work *Principles of Geology* was published between 1830 and 1833, and in the last volume he defined the Eocene on the basis that one to five per cent of its species still exist. The epoch lasted for 22 million years—between 56 million to 34 million years ago—and at the time of its beginning a great landmass existed where once the European Archipelago had spread. There were still plenty of islands about, including proto-Britain in the west and Iberia in the south, but stretching from the Turgai Strait in the east to Scandinavia in the north, a European proto-continent was beginning to take shape, one that no amount of rising seas or shifting tectonic plates has divided since.

For ten million years after the last dinosaurian jaws were stilled, the vegetation of Europe grew unchecked. Europe's forests had become as cathedral like—and yet more dense, gloomy and still—as the great forests of Borneo when first penetrated in the nineteenth century by the Italian explorer Odoardo Beccari. To him it seemed that the Bornean rainforests, the tallest on Earth, were places that had: 'remained untouched and unchanged since remote geological epochs, and where the vegetation has continued to flourish uninterruptedly for hundreds of centuries since the period when that land first emerged from the ocean.'[3] If we picture ourselves among the huge trunks, the gloom lit with luminescent insects and fungi, and a pervasive stillness and silence broken only by the odd, scurrying creature, we gain some idea of what the untrammelled forests of Europe were like.

Just prior to the great warming, a slight cooling caused sea levels to drop about 20 metres, opening a land bridge between Europe and North America, and allowing an American giant to enter Europe. *Coryphodon* was the largest creature to exist following the extinction of the dinosaurs. Descended from rat-sized North American ancestors of 10 million years earlier, *Coryphodon* belonged to an ancient, now-extinct order. Lumbering creatures, 2.5 metres in length and weighing up to 700 kilograms, with brains that weighed just 90 grams, they were probably rather unpleasant to look at—resembling overstuffed shrews.

Coryphodons were bulk-feeders on vegetation in the swampy woods of the New World, which then grew as far north as Greenland, functioning somewhat like dim-witted bulldozers and composters. Their impact, upon reaching Europe, was predictable. They found themselves in a larder that was 10 million years in the making. Too large to be attacked by any predator, and 'oversexed, overpaid, and over here' (to borrow a phrase), they feasted and wreaked havoc until eventually they exhausted their food supply.

As seedlings and understorey plants were eaten and older trees died without replacement, the timeless, umbrageous canopies of the ancient forests opened up, allowing sunlight to reach the forest floor—creating opportunity for lower-growing plants. Trails linking swamps and feeding places would have been trampled through the forest, with nutrients and seeds dumped along the way in piles of *Coryphodon* dung. With sunlight and a handy means of seed transport available, a far more varied canopy was established, and a greater variety of plants than ever before coexisted.

The *Coryphodon* invasion was just one event in a complex series of migrations that occurred around the dawn of the Eocene. We owe a great deal of our knowledge about these migrations to the work of

Dr Jerry Hooker. By the time I caught up with Jerry in June 2016 he
had been studying fossil mammals at the Natural History Museum in
London for more than 50 years. As he explained it, his work has involved
a lot of sieving. So much, in fact, that his hips had given out. Help,
however, was on the way—he was expecting a pair of titanium hips,
courtesy of the National Health Service. Given his sacrifices, I would
have thought that a gift of gold-plated ones would have been appropriate.

The sieving that palaeontologists like Jerry do is arduous, involving
agitating cumbersome sieves filled with sticky clay and sediment,
usually while standing in a freezing pond, in an effort to remove
the fine sediment and concentrate the fossils. After some time, all
that remains is rock fragments—and with any luck between three
and seven tiny teeth for every tonne of clay Jerry washes. Jerry will
sieve anything—from clay nearly 200 million years old to fresh stuff
formed just a few million years ago—as long as there's the prospect
of finding a fossil.

One of his finest moments was the discovery of the bones of the
world's oldest mole. He recovered the relics from 33–37-million-year-old
sediments on the Isle of Wight.[4] The creatures' teeth had been
described decades earlier, and while teeth can tell you what an animal
ate, they don't tell you whether it burrowed or bounded through the
bushes. Jerry persisted in working the site, washing its sediments
through very fine sieves until he had recovered minuscule foot and limb
bones whose spade-like features demonstrated that the creature was the
earliest known true burrower. The revelation opened the possibility
that moles first evolved in Europe, a view supported by DNA studies,
and the discovery in European rocks of fossils moles that live in North
America today.[5]

In my opinion, Jerry Hooker is both a national treasure and a saint.
Over his career he has found enough minute fossils to fill a few ciga-
rette packets. A mechanically minded friend, who had watched too

often as his mate bent over and sieved muck in a freezing pond, took mercy on him. After a bit of back-shed fiddling, he came up with a fossil 'washing machine'. I saw it whirring away in the yard at the Natural History Museum, shuddering and flushing muddy water as it concentrated the fossils. All Jerry had to do was put sediment in the top, and retrieve the concentrate from the bottom, which would be dried and then sorted through later. It's a tremendous piece of kit. Not as sophisticated as the Mars rover, but every bit as effective for exploring distant worlds.

Jerry's work has revealed that about 54 million years ago migrants poured into Europe from all points of the compass. From smallest to largest, the North American immigrants were ancestral shrews, squirrels, primitive ferrets, extinct otter-like creatures, pangolins, primitive carnivores and ancestral ungulates. From Africa came a modest contingent of primitive carnivores, while from Asia came the first even- and odd-toed ungulates, along with Europe's first primates and the ancestors of the modern carnivores.[6]

As a result of the arrival of these advanced mammalian lineages, the European fauna that had been evolving in isolation since the bolide impact was devastated. The frog-like Hainin beasts and their relatives, along with almost all the other mammals from Hainin, vanished. Extinction following invasion is of course a common occurrence in Earth's history and, indeed, it has happened repeatedly in Europe during the past 100 million years, but the European extinctions of 54 million years ago were exceptionally severe.

Among the victims were Europe's elephant shrews.[7] Elephant shrews are now found only in Africa, but the oldest African fossils do not appear until five million years after the first European ones. Elephant shrews are small, specialised creatures with noses like miniature elephant trunks. They feed mostly on insects, and they make pathways through vegetation, which they race along at high speed.

Some are credited with being the fastest mammals for their size on Earth. Oddly, they are one of the very few mammals, humans being another, to have a menstrual cycle.

Their unexpected presence in Europe offers a small diversion. Elephant shrews belong to a great division of mammals known as Afrotheria, which includes elephants, aardvarks, sea cows and a variety of lesser types. Afrotherians are so diverse in size and body shape that nobody suspected they were related until, in 1999, a DNA study revealed their affinities. But there were clues in their reproduction: all afrotherians have unusual placentas and produce more foetuses than can be nurtured in the womb.

It was long assumed that the afrotherians arose in Africa. But it seems odd that elephant shrews, alone among the Afrotheria, should have made the journey north into Europe at such an early stage. Alternatively, the afrotherians may have originated in Europe, and an elephant-shrew like creature crossed into Africa and gave rise to the great diversity of afrotherians—from elephants to golden moles—that inhabit the continent today. If so, then the Afrotheria are the sole survivors among the European mammals that evolved during Hainin times.

While Europe's mammals were devastated by the new invaders, its birds continued to thrive. As is to be expected of an island archipelago, there were many large, flightless species, among which was a two-metre-tall giant known as *Gastornis*.[8] The first fossils of the creature were discovered in the 1850s in the sediments of the Paris Basin by Gaston Planté, who went on to become a famous physicist, best known as the inventor of the lead-acid battery. So impressed was the palaeontologist Edmond Hébert by the 'studious young man full of zeal' who turned up at the Paris museum with his finds that he named the creature in Gaston's honour.

Gastornis evolved in Europe from goose-like ancestors that had become flightless in their island environment. When the land bridge to North America opened, *Gastornis* crossed onto the continent, and the recent discovery of fossils in China indicate that they reached Asia as well.* *Gastornis* had a massive beak capable of crushing hard objects, and generations of palaeontologists believed them to be predatory: many older illustrations show the great birds catching and consuming early horses. But a recent analysis of calcium isotopes indicate that *Gastornis* was exclusively herbivorous.[9] By 45 million years ago, these gigantic birds had become extinct in North America and Asia, and then subsequently vanished from their last stronghold—their ancestral European homeland.

Modern skinks and more amphisbaenids arrived.[10] Meanwhile common frogs and toads came and went. The true toads had arrived in Europe about 60 million years ago (presumably from Asia) only to disappear, before recolonising about 25 million years ago. Beginning around 34 million years ago, the green frogs (Ranidae) arrived, perhaps from Asia or Africa.[11] Flittering in at around this time, from parts unknown, came Europe's first bat.[12] It is astonishing that bats appear to have been absent from Europe, Asia and North America before this. So where did they come from? The world's oldest bat fossils are found in Australia, but no possible bat ancestors or near relatives are known from that continent. The origin and spread of the bats remains one of the greatest mysteries of palaeontology.

Jerry Hooker's work revealed that 54 million years ago, a second migration occurred, just 200,000 years after the first. The great warming caused the sea to rise 60–80 metres during just 13,000 years, severing the land bridges to Asia and Africa. But due to volcanic activity, the land bridge to North America remained open, and marsupials, early primate-like creatures and some primitive carnivores used it to reach Europe. At the same time, something unprecedented

---

* In North America, it was long known as Diatryma.

happened: European creatures, including the ancestors of the dogs, horses and camels, all of which had arrived in Europe from Asia just 200,000 years earlier, made a mass migration into North America.

In a sense, this great migration laid the foundations of the modern world, for it enabled the evolution, in North America, of horses, camels and dogs which, in our hands, would help transform our planet. It also foreshadowed Europe's future: the biological wealth of Asia had been poured into the European proto-continent, and then a way to the Americas was discovered.

CHAPTER 10

Messel—a Window into the Past

Thanks to one of the world's most extraordinary fossil deposits we know more about life on the proto-continent of Europe a few million years after the great warming than any earlier period. The deposit, formed 47 million years ago, is exposed in an old lignite mine at Messel, near Frankfurt in Germany. Fossils from Messel can look like the remains of animals that have been pressed between the pages of a book, with impressions of hair, skin, and even stomach contents often present. This is about as far from the solitary teeth that people like Jerry Hooker study as imaginable, making the Messel fossils immensely valuable.

Marvellous fossils were being discovered at Messel as early as 1900, but in the 1970s the burghers of the town proposed that the site be used for landfill. Not since Pope Sixtus V suggested turning Rome's Colosseum into a wool factory to provide employment for the city's prostitutes (a fate avoided only by the pontiff's premature death) have the heritage values of Europe been so ignored. The authorities came to their senses in 1991 and purchased the pit to secure scientific access. Between 1971 and 1995, however, amateur collectors had had free access to the priceless fossils, and therein hangs a tale of human frailty and greed that chills the bones of palaeontologists.

On 14 May 2009, a press release headed 'World Renowned Scientists Reveal a Revolutionary Scientific Find that Will Change Everything' was received at news offices around the globe.[1] At a press conference held the following day, at the Natural History Museum in New York, it was claimed that a missing link in human evolution had been found at the Messel pit, a treasure that compared in heritage value with the Mona Lisa. The research team making the claims was led by Jørn Hurum of the University of Oslo's Natural History Museum. He bestowed the nickname Ida (after his teenage daughter) on the fossil. Hurum claimed that, 'This specimen is like finding the lost ark... It is the scientific equivalent of the holy grail'.[2] The fossil 'missing link', it transpired, was an exquisitely preserved 58-centimetre-long skeleton of a small primate, surrounded by traces of fur and replete with its last meal. In a scientific paper published two days later, researchers claimed that the small creature, which they named *Darwinius masillae**, was an intermediate form between the more primitive primates known as prosimians and the simians, the group that includes monkeys and humans. If this was indeed true, it would rewrite our understanding of early primate evolution. Until *Darwinius* came along, it was generally believed that simians derived from tarsier-like creatures.

Scientists do not like spectacular claims being made in the popular press, especially if they are announced before the supporting evidence is published in a reviewed journal. A newspaper headline 'Origin of the Specious', published shortly after the announcement, should have warned Hurum and his coauthors of what was to come.[3] Nils Christian Stensteth, one of Norway's leading biologists, called the claims 'an exaggerated hoax' that acted to 'fundamentally violate scientific principles and ethics'.[4] Moreover, analysis showed that Hurum's team was wrong.

---

* The first name in celebration of the bicentenary of the birth of Charles Darwin, the second from the Roman name for Messel.

Ida is not on the human lineage, but is an early primate known as an Adapiform, which is similar to lemurs.

The specimen had been unearthed by an amateur fossicker in the Messel pit in 1983. Because of the way fossils are preserved at Messel, the skeleton came in two parts—a slab containing the bones (a 'positive', if you like) and a counterpart containing impressions (the 'negative'). The negative surfaced at a private museum in Wyoming, USA, in 1991, but was soon shown to have been partly faked: it was a composite, made up of the remains of two different creatures.* In 2006, the positive impression was offered to Hurum for US $1 million. He bought it for $750,000, a price that would put pressure on most museum budgets. And with financial pressure comes the need to maximise publicity and significance. A contract was signed for a popular book, and the History Channel reportedly paid more for the story than they had for any other program.[5] Unregulated amateur digging at sites like Messel and paying of vast amounts for fossils can create a toxic hazard for researchers. Had the good burghers of Messel realised in 1971 what a treasure they had in the old lignite pit, and immediately protected it, the farce might have been avoided.

The Messel deposits formed 54 million years ago at a time when the descendants of the creatures that had reached the European proto-continent were diversifying and adapting to local conditions. Among them were the ancient palaeotheriids, early relatives of the rhinos, tapirs and horses. A variety of strange and primitive ungulates belonging to six artiodactyl families, including the duiker-like anoplotheriids and the rabbit-sized dichobunids, also flourished. All of these families were unique to Europe, and all were small creatures.[6] Like Nopcsa's

---

* It is not known how much was paid for the cleverly tricked-up specimen, but I suspect it was a great deal.

dwarfish dinosaurs, Europe's Eocene mammals had adapted to life on a tropical island by reducing in size.

At that time, Germany was located 10° south of its present location, and was a volcanic, tectonically unstable place. The Messel pit was then a lake surrounded by luxuriant rainforest, at the bottom of which, in an oxygen-free environment, formed the lignite and oily shale later mined there. The nearby volcanoes made the lake a perfect future fossil site. They would occasionally belch CO_2, which, being heavier than air, would sink to the lowest point in the landscape—the lake surface—and hang there. Any bird or bat flying over the lake, or creature coming down to drink, would lose consciousness and sink to the bottom, where the anoxic chemistry of the sediments would prepare it for eternity as expertly as any mummy-maker.

In some of the Messel fossils, there is enough detail to give the impression of an old black-and-white photograph of a vanished creature. But even colour is preserved in some small beasties, such as jewel beetles. And sometimes the fossils bring the ecological fabric of the forest to life: the impressions of an ants' jaws on a leaf fragment led researchers to surmise that the ant had been afflicted with a parasitic fungus that alters behaviour, urging its host to climb to a high place and hang there until it is dead, so that the fungus can then release its spores into the breeze.

Among the most extraordinary of Messel's treasures are nine mating pairs of pig-nosed turtles (a creature found in the dinosaur-age Charentes deposits). As a student of the fossil record, I can assure you that it's not often that creatures are transformed, in *flagrante delicto*, into *memento mori*. Among the many mammals from Messel is the tapir-like *Propaleotherium*. Bodies of the 10-kilogram creatures have been found complete with near-term foetuses, and the contents of their last meal (berries and leaves) in their stomachs. There are also some surprises such as *Eurotamandua*, a scaleless pangolin that looks astonishingly like a South American anteater. But it is Messel's birds that constitute the true treasure of the site. For birds, lacking teeth, make poor fossils and are difficult to classify from other fragments. At Messel an entire avifauna is preserved, as if in aspic.

Some of the Messel birds—including falcons, hoopoes, an owl, ibises and an ancestral pheasant-like creature—are to be expected. But others are out of place, unexpected or downright bizarre. Among those out of place are a kind of potoo (a nocturnal bird like a nightjar), a hummingbird, a sunbittern and a relative of the carnivorous cariamas, all of which today inhabit South America, but not Europe. A primitive ostrich-like creature and a mouse bird (today, exclusively African) also join this group. Among the unexpected is a kind of gannet that hunted over freshwater, while among the bizarre must surely be counted a parrot that lacked a parrot-like bill, and a strange creature that looked like a mixture between a hawk and an owl, but which had membranous, ribbon-like breast feathers.[7] What is missing from Messel and indeed all of Europe at this time are ancestral larks, thrushes, orioles and crows, all perching birds, despite the fact that the greatest part of Europe's avifauna today consists of perching birds.

What to make of the high proportion of South American bird species at Messel? Oddly, there is good geological evidence that at the time South America, though it lay close to Africa, was entirely isolated by water, so over-water migration was the only possible route. As things stand, we have no convincing explanation as to why so many birds that are restricted today to South America flourished in Eocene Europe.[8]

CHAPTER 11

The European
Great Coral Reef

It's 1 June 2016 and I'm standing in front of a grey cabinet that holds the Natural History Museum's collection of fossil corals, hardly believing what I'm seeing. It looks like an irregular lump of rock, but Brian Rosen, one of the museum's coral researchers, explains that it is in fact the holotype (the name bearer for the species) of *Acropora britannica*—a member of the great *Acropora*, or staghorn group of corals, which today are among the most important of the reef-building corals. Named by Dr Carden Wallace, an Australian expert on this group of reef-forming corals (the Acroporidae), it was found in 37-million-year-old (latest Eocene) sediments near the picturesque village of Brockenhurst in the New Forest, near Southampton.

The rocks around Brockenhurst have yielded fragments of an extraordinary marine fauna, including *Acropora anglica* and *britannica*—two species that are among the earliest members of the two great branching coral species groups—'robusta' and 'humilis II'—which make up most of the ocean-front corals in the Indo-Pacific today.[1] Could Brockenhurst in the New Forest really have been the birthplace of Earth's magnificent coral reefs? For more than a century, geologists have known that 37 million years ago the Brockenhurst area was where the coast of the proto-European continent faced the open Atlantic

Ocean. There, according to one nineteenth-century geologist, 'coral reefs exposed to furious surf and the wash of a great ocean' had formed a bulwark against the southerly wind and swells.[2]

Recent researchers doubt that an actual coral reef occurred in the Brockenhurst area, though reef-forming corals clearly grew there, and fast-growing branching corals such as *Acropora* thrive in such energetic environments. Moreover, Brockenhurst does not mark the beginning of the *Acropora* genus, for there are a few older fossils from France, and a single record from Somalia, dating to around 55 million years ago. The Brockenhurst fossils, however, do form part of the evidence that many modern coral reef organisms originated in the European section of the Tethys Sea.

Thanks to a most exceptional fossil deposit in Italy, we know a little about the animal communities that thrived when the *Acropora* corals first evolved. For more than 400 years, travellers have been visiting Monte Bolca near Verona to peer into a 50-million-year-old fishbowl, or *Cava della Pesciara*, as the Italians call it. The very earliest written record of a visit to the site, by Pietro Andrea Mattioli, dates to 1554: 'some slabs of stone which, on being split in half, revealed the shapes of various species of fish, every detail of which had been transformed into stone'.[3] Over the years noblemen, cardinals and even the emperor Franz Joseph passed through, and left with fossil fish mementos.

About 50 million years ago, when the fossils were living creatures, the rocks of Monte Bolca were forming in the Tethys Sea. The fish and other creatures preserved in the deposit appear to have inhabited a lagoon that formed between the land and a reef (though modern reef-building corals such as *Acropora* have not been found there). Nearby, the remains of crocodiles, turtles, insects and plants have been found, all exquisitely preserved. Among the plants are coconuts and other palms, figs and eucalypts.[4] The fish fossils are some of the most spectacular and beautiful found anywhere on Earth: some look as if they are still swimming and retain traces of patterning and colour.[5]

The marvellous preservation of the fossil fish from the *Pesciara* cannot be fully explained. The best theory to date is that occasionally a toxic algal bloom broke out that killed the fish en masse and their bodies drifted down to oxygenless water in the lagoon's deeper parts. Whatever the cause, about 250 species of fish are represented in the deposit. And we would have none of them except for an unlikely geological event. The entire region around Verona was volcanic and highly unstable at the time the layers formed. Before it had hardened to rock, the slab of sediment containing the fish, which is several hundred metres long and 19 metres thick, was transported, intact, a considerable distance—perhaps by an underwater landslide.

The most important thing about the Monte Bolca fauna is that it is the oldest known occurrence of the community of fish that inhabit coral reefs today. Despite the presence of a few now-extinct fish families, the 250 species preserved at the site are broadly similar in type and form to kinds that can still be seen on the world's reefs, including rays, angelfish and eels. But both butterfly fish and parrotfish, which abound on modern coral reefs, are absent from the deposit, suggesting that they probably evolved later.[6] An astonishing exception, however, is a single fossil of a handfish, so-named because they 'walk' on fin-like hands. Today they occur only in the cold waters of southern Australia.[*] Some years ago I had a choice of visiting the Galleria dell'Accademia in Florence to see Michelangelo's David, or going to the natural history museum in Verona to see the fossil fish. You can guess which I chose. I arrived in Verona on a sunny Thursday and made a beeline for the

---

[*] The Monte Bolca handfish is known as *Histionotophorus bassani*. Of the 14 living species, 11 are restricted to Tasmania. In the Eocene, they must have existed along the length of the Tethys.

museum, which is located across the river from the city centre, and was dismayed to discover that it was closed without notice. In a tale that I'm sure is familiar to many museum goers in Italy, I returned the following day only to find that the museum was closed from Friday to Tuesday each week, the very day I was booked to leave the city! My only consolation was roaming Verona's well preserved Roman arena, where some of the tiered seats contain the remains of ammonites the size of truck tyres, their surfaces polished to a gem-like finish by the posteriors of ancient Romans. Did they, I wonder, ever ask themselves what those great round shell-like shapes were doing there in their rocky seats?

CHAPTER 12

Tales from the Sewers of Paris

Around the time the Monte Bolcan fish were breathing their last, a region in northern France was a languorously warm embayment of the Atlantic Ocean. The sediments that fell to the bottom of that sea are now known as the rocks of the Paris Basin, and in 1883 the French geologist Albert de Lapparent—who is perhaps best known for his efforts to link Britain with the rest of Europe by a rail tunnel—coined the name Lutetian (after the Roman name for Paris) for the geological age during which the rocks of the basin were formed.

The Paris Basin rocks include the famous Paris stone—a limestone that has been used for construction since Roman times—and whose warm creamy-grey colours give the city an unmistakeable beauty. As I wander its streets, it's not just scenes of the French Revolution that fill my mind, or the delicious smells of fresh bread and cheeses that beguile me, but traces of that long-ago Paris—a place of marine giants and tropical creatures—and a wondrous biodiversity.

There is no better place to see the traces of Paris's lost glory than at the *Muséum Nationale d'Histoire Naturelle* in the Jardin des Plantes. One of the world's oldest museums, it's where both le Comte de Buffon and Georges Cuvier (the father of palaeontology) spent their working lives. During the opening decades of the nineteenth century,

Cuvier laid down a number of 'doctrines', some of which survived the test of time better than others. He was right in arguing that extinction did occur (a fact doubted in his day), but wrong in arguing against evolution.* Instead, he developed the idea that catastrophes had periodically extirpated life, and each time God had created life anew. This was a logical consequence of his reading of the fossil record.** As Cuvier saw it, most fossil species remain similar in form from their first occurrence to their last, and 'missing links' are extremely rare. This was also known to Darwin, and it worried him exceedingly. But Darwin perceived what Cuvier did not: that prehistory is so vast that fossils give us nothing but the tiniest glimpse into the life of times past. As Signor-Lipps point out, this means we almost never see the origin of a species or its final extinction in the fossil record.

Some of Cuvier's most enduring work was undertaken with Alexandre Brogniart, a teacher at the Paris Mining School. Together they examined fossils that had been unearthed around the city, many of which were encountered during the excavation of Paris' famous sewers. Another prolific area for discovery was Montmartre, where mining for gypsum to make plaster of Paris almost undermined the famous hill.*** It was the abundance of fossils, from both terrestrial and marine environments, preserved there that allowed Cuvier to work out the rules of geological succession (that younger rocks overlie older ones).

---

* As remarkable as it now seems, the idea of extinction was opposed on the basis that supposedly extinct species surely survived at some location—perhaps in the unexplored American West—and on the theological grounds that God would not extinguish his own creations.

** A much easier argument to make prior to the publication of Darwin's *Origin*. Cuvier died in 1832.

*** The gypsum used to create the famous plaster was formed around 50 million years ago, when vast lagoons dried out, leaving the mineral in thick deposits.

Despite a long-term global cooling trend, conditions in the shallow seas around what was to become Paris remained favourable for the growth of marine organisms.[1] One beneficiary was the giant bell-clapper shell (*Campanile giganteum*) described in 1804 by Jean-Baptiste Lamarck.* Perhaps exceeding a metre in length, it was the largest gastropod ever to exist, and its remains, which are largely restricted to the Paris Basin, were frequently encountered during the excavation of the sewers. A single species of bell-clapper shell survives today— in rocky habitats in the cool, shallow, waters off southwest Western Australia. While reaching only a quarter the length of its gigantic European relative, it is a rare and wondrous reminder of the glories that once swarmed the sea where Paris stands today.

But what of life elsewhere in the Tethys, that marvellous lost sea that bathed the proto-continent in salty, balmy warmth? Another true giant was the largest cowrie that ever lived, *Gisortia gigantea*. Its exquisitely fossilised shells, the size of rugby balls, date back 49–34 million years. They have been found in places as widespread as Bulgaria, Egypt and Romania. Cowries, with their porcelain-like lustre, are among the most beautiful of all gastropods. Sadly, nothing even remotely similar in size to the great *Gisortia* survives in today's oceans.

The Tethys was also the headquarters of the mighty *Nummulites*, a few species of which survive in the Pacific today. The name is derived from a Latin word meaning little coin, and these single-celled organisms abounded during the Eocene. *Nummulites* creep along the bottom of the sea, feeding on detritus, and laying down disc-like, many-chambered internal shells of calcium. The Tethys provided a perfect habitat for them: tropical, shallow and sunlit. In Turkey, fossil *Nummulites* as large as 16 centimetres in diameter have been found.

---

* Lamarck is best known for his theory of evolution, in which he posited that a creature's experience during its lifetime could be passed on to its offspring.

These giants are estimated to have lived for a century, making them the longest-lived single-celled organisms known.[2]

Such was the abundance of *Nummulites* throughout the length of the Tethys that in many places their remains formed a distinctive kind of rock called nummulitic limestone which, since ancient times, has been much favoured for construction. The origin of this ubiquitous rock—it was used by the ancient Egyptians for construction of the pyramids—was long a source of mystery. Herodotus spread one early misconception: that the *Nummulites* were the petrified remains of lentils which the Egyptians fed to their slaves as they laboured over the mighty structures. But even in the early twentieth century, the presence of *Nummulites* in the pyramids continued to befuddle, as the sad tale of Randolph Kirkpatrick, assistant keeper of lower invertebrates at what is now the Natural History Museum, London, illustrates.

One of the greatest battles in geological science was that waged between the Plutonists and Neptunists on the origins of Earth's surface. The Plutonists, who had Thomas Huxley on their side, asserted that rocks such as basalt and granite, which originated in a molten state deep within the Earth, were the primary source, and that the other rock types such as sandstone and slate were formed from their breakdown and re-deposition as silt and mud. The opposing Neptunists, who counted Goethe among their number, believed that the Earth was originally covered in ocean, and that all rocks originated as deposits on the floors of ancient seas. By the mid-nineteenth century the matter had been all but settled in favour of the Plutonists. But then in 1912, Kirkpatrick, dropped a bombshell that reignited the debate.

It had not escaped Kirkpatrick's notice that the pyramids were almost entirely composed of *Nummulites*. As he scoured rocks for evidence of yet more *Nummulites*, he began seeing them in every rock type he placed under his microscope. In his great opus, *The Nummulosphere* (which opens with a stupendous frontispiece featuring

Neptune driving a *quadriga* over a watery globe), Kirkpatrick used this supposed ubiquity of *Nummulites* to revive the theory of the Neptunists, arguing that the entire crust of the planet, and ultimately the Solar System and the Universe, consisted of the fossilised fragments of *Nummulites* that had lived in a primal sea.[3]

Historians of science have often wondered how a staid and doubtless sober curator at one of the world's most august natural history institutions could go from publishing serious and important research to making such outrageous claims. When I have discussed the question with experts on corals, they tell me that a life spent researching the complex biology of organisms like corals and sponges can alter a man. George Matthai worked at the Natural History Museum shortly after Kirkpatrick. After describing countless new species of corals, including many of those that make up Australia's Great Barrier Reef, he suicided.

Matthai's colleague Cyril Crossland also suffered for the cause. In 1938, after decades of strenuous work studying corals in British, Egyptian and other research institutions, he took up a position at the University of Denmark's Zoological Museum. Perhaps extreme dedication to his research saw him eschew the dangers emerging to the south, or perhaps his deafness left him unaware. Prior to his death in 1943 he was seen riding Copenhagen's trams, roundly abusing the Nazis in a cultivated English accent. The heroic if rash Crossland was sadly missed by his colleagues, who named sixty species of marine organisms after him.

Apart from his obsession with *Nummulites*, Kirkpatrick exhibited no signs of mental infirmity. He was sincere in his beliefs about the Nummulosphere and he published images that he claimed depicted the remains of *Nummulites* in basalts, granites and meteorites—rocks in which no fossils have ever been found—so that others could verify his claims. My son David, who is also a scientist, on hearing Kirkpatrick's story, remarked to me that many a researcher, after spending thousands of hours peering down a microscope at some repeated shape, begins to see it repeated *ad nauseam* on blank walls, distant vistas, even a spouse's face. And it's not just images but theories that can get imprinted and reflected, causing a scientist to see evidence for their favourite theory everywhere. Perhaps the affliction should be known as nummulitis.

As Kirkpatrick worked, Otto Hahn—an intensely patriotic German lawyer and amateur petrologist turned Swedenborgian who believed life originated in outer space—was spending long hours staring down his microscope at what he took to be the fossilised remnants of algae. Hahn, like Kirkpatrick, was a Neptunist, but he thought the idea that the Earth's rocks were made of *Nummulites* was ridiculous. He proposed that they were composed of a fossilised forest of algae which had originated on meteorites. He also 'discovered' the fossil of a minute, triple-jawed, algae-eating worm, which he named *Titanus bismarcki*, in honour of the German chancellor. Bismarck had other things on his mind, for the European powers had embarked on the Great War.

By 49 million years ago, the continued growth of the proto-continent of Europe was profoundly altering the seaways surrounding it. To the south, the Tethys was narrowing, as was the Turgai Strait, separating Europe from Asia. Except for a recently formed and still narrow north Atlantic, the ever-narrowing Turgai was the only connection between the waters of the Arctic Sea and the rest of the world's oceans.

The Arctic Sea has not always been frigid and ice-covered. Forty-nine million years ago it was more like the Black Sea of today—with its oxygenless and very deep salty layer beneath fresher waters—though the Arctic Sea was then more tropical than today's Black Sea. This was also a time of intense rainfall, and as the Arctic Sea became more cut off from the rest of the ocean, the runoff from rivers began to pool in its upper layers, which freshened to the point that a particular kind of weed, known as *Azolla*, could grow there.

If you have ever had a pond you will know *Azolla*, a.k.a. pondweed, duckweed fern or fairy moss. Its tiny, crinkly leaves often first appear as a minute speck of floating green that seems to increase slowly. But by the time it has covered 10 per cent of the pond's surface, it's only days away from a complete takeover. Given warmth and the right nutrients, *Azolla* can double its mass every three to ten days.

The evidence that *Azolla* once grew in the Arctic Sea is today buried beneath thousands of metres of frigid sediments and water below a skin of ice. It may have lain unrecognised forever were it not for some very expensive cores punched deep into the Arctic sediments in 2004 by drill-rig crews searching for oil. The last thing they expected to find was evidence of pondweed. But there it was—in layers of varying thickness distributed through at least eight vertical metres of sediment. The fossils were soon dubbed *Azolla arctica*.[4] The presence of *Azolla* has now been confirmed in more than 100 drill cores taken from throughout the Arctic region, with the greatest concentrations being in cores drilled from the Arctic Sea itself.

At least five species of *Azolla* were growing in and around the Arctic Sea 49 million years ago.[5] Warmth, fresh water and the nutrients brought in from rivers provided all that the weeds required. At its height, the *Azolla* bloom covered about 30 million square kilometres of ocean—an area the size of Africa.[6] The weed grew so vigorously, sucking in atmospheric CO_2 in the process, that it reduced the global atmospheric concentration of CO_2 from at least 1000 parts per million to 650. And all that captured carbon would go on to form the Arctic oil reserves that the petro-giants are so keen to get at today.

The *Azolla* blooms eventually extinguished themselves, for the lack of CO_2 lowered global temperatures so substantially that rainfall declined at the poles, causing inflows of freshwater and nutrients to taper off, which starved the weed.* As temperatures continued to drop,

---

* This is an excellent example of Gaia's self-regulation: a negative feedback loop that prevents life from pushing Earth's climate out of the habitable zone.

a layer of ice formed over the Arctic Sea. Thus was a new icehouse world initiated by a minute weed. Initially, however, the lowering of CO_2 concentrations had remarkably little effect on Europe—it was almost as if the preconditions for a major change had been established, but the trigger had yet to be pulled.

II

BECOMING
CONTINENTAL

34–2.6 Million Years Ago

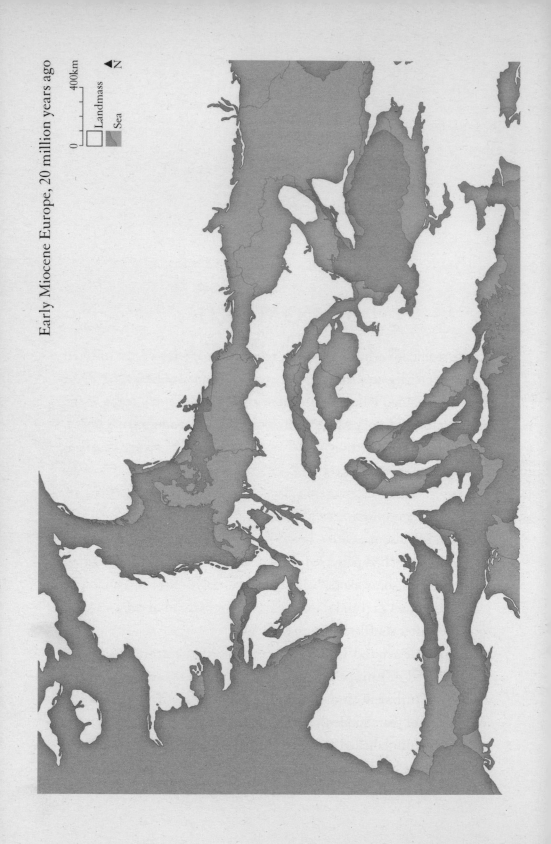

Early Miocene Europe, 20 million years ago

0   400km

Landmass
Sea

N

CHAPTER 13

La Grande Coupure

As the twentieth century dawned, bringing such wonders as powered flight, electricity and the motor car, the Swiss palaeontologist Hans Stehlin remained glued to his microscope, sitting in his office at the Basel Natural History Museum in Switzerland, pondering old bones. He had become something of a legend for his dogged pursuit of palae-ontology, but it seems there was more to his dedication than scientific interest. According to museum folklore he had been thwarted in love, and to forget his misfortune had poured all his energy and passion into his work. Handsome, with a Freud-like beard and piercing eyes, it was also said that he had perfected the death stare. Whenever he required the skeleton of some exotic beast to compare with his fossil bones, he would visit Basel Zoo and stare at the appropriate animal, which would soon thereafter shuffle off its mortal coil.

Sometime around 1910 Stehlin realised that a dramatic change had occurred in Europe's fauna about 34 million years ago. During a period of turbulent climate change, many species that had survived for millions of years suddenly became extinct, and a host of new species arrived. Stehlin labelled the event *la grande coupure*—the great cut. And scientists have been arguing about its precise causes and timing ever since. *La grande coupure*, now dated at 34 million years ago, has been widely hailed as marking the end of that inordinately long tropical epoch, the Eocene, and the beginning of the colder, drier Oligocene. In general terms this is true, for *la grande coupure* marks a fundamental reorganisation of climate—from a predominantly greenhouse world to an icehouse.[*]

The cause of the climatic shift this time appears to have been the parting of South America from West Antarctica. The Drake Passage, as the seaway separating these landmasses is known, was initially shallow and remained so for millions of years, but water flow was sufficient to allow the establishment of an ocean current that encircled Antarctica. This allowed cold water to build up and an ice cap to form, which led to a fundamental reorganisation of ocean currents and winds, bringing much cooler conditions.

In Europe, this shift was accompanied by changes in the hydrological cycle. The story of precisely what happened is eloquently told in snail shells—specifically in the fossilised shells of the freshwater snail *Viviparus lentus*, which once thrived in a coastal floodplain where the Solent now lies, off the Isle of Wight.[1] Lister's river snail (*Viviparus contectus*)—a large freshwater species with a striped shell, which survives in lakes in Britain today—gives us a good idea of how its ancient relative looked and lived. Isotopic studies of the fossilised snail shells reveal that cold water exported into the north Atlantic from Antarctica caused air temperatures in southern Britain to drop by 4–6° Celsius. But in summer, which is when the snails grew, temperatures plummeted by nearly 10° Celsius. Along with the changing climate, something else very important was occurring. The Turgai Strait, which stretched from the Tethys Sea via what is now the Caspian Sea and on to the Arctic

---

[*] The Eocene's end is, in fact, marked by the demise of the obscure single-celled planktonic foraminiferal family known as the Hantkeninidae. But *la grande coupure* occurred in two pulses about 350,000 years apart, just after the last hantkeninids found their eternal rest.

Sea, would in parts disappear. As a result, Europe and Asia were finally connected. And, at about the same time, Europe and North America were joined by a land bridge for one final, brief, moment.

In 2004 the redoubtable Jerry Hooker and his colleagues took another look at Stehlin's big cut. Examining sediments exposed around the Solent estuary and in northern France and Belgium, they showed that things were, as usually proves to be the case, far more complex than they had first appeared. There was among the fossils they examined evidence for two quite separate extinction events—a smaller one, which coincided with a change in climate, and a larger one a few hundred thousand years later, which coincided with the arrival of new mammalian invaders.[2]

One of the few surviving lineages was the dormouse. Dormice are not actually mice but members of the family Gliridae, ancient rodents whose ancestors arrived in Europe 55 million years ago from North America. They proliferated, adapting to European conditions and expanding into a wide variety of ecological niches. For more than 40 million years they remained restricted to Europe, before spreading to Africa some time after 23 million years ago, and only much later to Asia. They are Europe's oldest and most venerable mammals, though their current diversity is the merest vestige of their former glory.

The Oligocene extended from the *la grande coupure* of 34 million years ago to about 23 million years ago. Despite the cooler conditions, much of Europe's vegetation remained largely subtropical to tropical. The Turkish coast abounded in mangroves, nipa palms (which cannot tolerate temperatures below 20° Celsius) and other vegetation today associated with tropical southeast Asia.[3] About 28 million years ago sea levels dropped once more, and even cooler conditions prevailed. Yet the Turkish plant fossils indicate that rattans and cycads persisted distant from the sea in a forest that was largely made up of flowering plants, including *Engelhardia* (today restricted to southeast Asia), hickory (no longer found in Europe), and an ancestral hornbeam.[4] Though forests continued to hold sway over much of the land, deserts and grasslands were gaining footholds in places like Iberia, allowing for a greater diversity of animal species.

Meanwhile, the land itself was undergoing momentous changes, not the least of which was the uplifting of Europe's most majestic mountain range, the Alps. The origins of the Alps go back to the age of dinosaurs, but that early uplift was followed by a period of quiescence, which ended at the beginning of the Oligocene when a downwards bending portion of the European plate broke off and started to make its way towards the surface, forcing the modern Alps to rise.

The development of Europe's current topography involved much subsequent folding, faulting and thrusting. Fragments of land were moving in all sorts of ways and in different directions. Some appear to have shot from one side of the Mediterranean to the other with (by geological standards) great alacrity. Others were shoved deep into Earth's mantle, becoming melted or deformed in the process, and sheets of rock that originated in Africa, known as nappes, were pushed over rocks of European or oceanic origin. One such piece of African geology was destined to become the peak of the Matterhorn, often called the 'African mountain'.[*] As Africa and Europe closed in on each other, vast expanses of the ancient Tethys Sea floor were uplifted then eroded away, giving rise to some of the spectacular limestone land-scapes that can be seen today in the foothills of the Alps.

The driving force was Africa. Originally drifting north-northeast, from about 16 million to seven million years ago, it swerved slightly in

---

[*] To be precise, it is the portion of the Mattterhorn lying above 3400 metres that is African. The lower slopes are composed of marine sediments or European rocks. The African rock is part of the Apulian Plate, itself originally part of Africa, most of which underlies the Adriatic Sea. It was thrust above the other rocks of the Alps as the Apulian and Eurasian plates collided.

a north-northwesterly direction. Thereafter it shifted again and began moving northwest—the direction it continues in today.[5] This counter-clockwise twisting severed the Tethys Sea, temporarily blocked up the Strait of Gibraltar, and thrust the Alps skywards. Indeed, as Africa makes its way north, the Alps continue to rise—at the rate of between one millimetre and one centimetre per year, though they are weathering away almost as fast as they grow.

I wouldn't be surprised if George Orwell took inspiration from the Oligocene for his novel *Animal Farm*. Whatever the case, as in Orwell's cautionary tale, the Oligocene's signature species were an unsavoury group of pigs and pig-like creatures, foremost among which were the entelodonts, more popularly known as hell pigs, or terminator pigs. The ancestors of these cow-sized creatures had migrated from Asia about 37 million years ago. Palaeontologists will tell you that they were not pigs but relatives of hippos and whales. But had you met one, your first impression would have been of a gigantic, hyper-carnivorous warthog.

Perhaps the entelodonts' most unlovely feature was their oversized heads. They were garishly ornamented with bony warts the size and shape of a human penis, and their almost crocodilian jaws bore savage tusks and grinding molars. Unlike modern pigs, which aren't averse to flesh-eating but mostly subsist on vegetable matter, the entelodonts were apex carnivores. And they were fast, their long, slender legs carrying their 400 or more kilograms with a speed that would leave wild boars, and humans, in the dust.

Their efficiency as carnivores is attested to by the discovery of caches of fossilised remains of their victims. One North American cache consisted of the skeletons of several sheep-sized ancient camels.[6] It is thought that the entelodonts ran down herds of smaller beasts and slashed and battered them en masse, a practice that compelled them to store their leftovers. Or perhaps rotting flesh was easier on their guts,

There were also many rodents in Europe during the Oligocene, including a proliferation of dormice, squirrels, voles and beavers. One small creature of note is the ancestor of the desmans. Europe's most distinctive mammals and members of the mole family that have taken to life in rivers, streams and ponds, there are only two species—one found in the Pyrenees and the other in eastern Russia. The Russian desman, which can weigh half a kilogram, is by far the largest member of the mole family—large enough to be sought for its fur.

CHAPTER 14

Cats, Birds and Olms

About 25 million years ago, as the Oligocene was nearing its end, the first member of the cat family, known as *Proailurus,* emerged from Asia and entered Europe. Roughly the size of a domestic cat, its fossils have been found in Germany, Spain and Mongolia. For 10 million years after they arrived in Europe, despite the great numbers of small rodents present, cats showed no sign of prospering, let alone attaining the dominance among ambush predators they have today.

Bird bones make miserable fossils, and the incomplete record from the Oligocene in Europe leaves much room for conjecture. A few scraps found in England and France have been trumpeted as evidence that New World vultures (a group that includes the condors) once soared in Europe's skies. But a reassessment suggests that they are the bones of an Oligocene cuckoo-roller.[1] The bones of *Paracygnopterus* have been found in southern England, which some claim to be the oldest member of that most ornamental group of waterfowl, the swans. Others, however, opine that it is merely a goose.[2]

The evidence is better that loons, aquatic diving birds, abounded on lakes in Oligocene England and Belgium, chasing fish in much the same manner as existing species, though whether they uttered anything like the haunting cry of their surviving relative the great northern

diver cannot be known. Less expectedly, a species of secretary bird—
a creature familiar today on the savannas of Africa—stalked the grasslands of France. But the most important Oligocene arrival was that
of Europe's first songbirds—perhaps at the time of *la grande coupure*.
These early migrants subsequently became extinct in Europe, replaced
by later songbird migrants.[3]

Recent DNA research shows that songbirds, parrots and falcons
are related, and that this highly successful group originated around the
time of the dinosaur extinction in the Australian section of Gondwana.
That falcons and robins are more closely related to each other than are
falcons and hawks seems nonsensical. But the avian body plan is highly
restricted by the requirements of flight, so convergent evolution, in
which unrelated creatures develop similar characteristics, is common.

The songbirds are by far the largest and most successful birds.
Their 5000 species, divided between 40 orders, comprise 47 per cent of
all bird species. Eighteen of Britain's most abundant avian species are
songbirds, as is the most abundant wild bird on Earth, Africa's red-
billed weaver bird, of which 1.5 billion are thought to exist. That said,
the great majority of songbirds fall into just one order, the perching
birds, or Passeriformes, which take their name from the Latin word for
the sparrow. All of the little birds that forage among leaves are perching
birds, as are crows and magpies—and one thing that sets them apart is
the possession of a hind toe operated by an independent set of tendons.

The first inklings about the origin of the songbirds came in the
early 1970s. Charles Sibley, an ornithologist working at the University
of California, discovered that if he purified and boiled double-stranded
bird DNA, the strands would recombine when the mixture cooled.
If he mixed the DNA of closely related species, the bond between
the strands was stronger than for more distantly related pairs.[4] Since
Sibley's day, genetic studies have become immensely more sophisticated.
In 2002 it was demonstrated that New Zealand's wrens sit at the base

of the songbird family tree. Other studies show that the second oldest branch of the songbird family tree includes Australia's lyrebirds and scrub birds, while the third oldest includes Australia's tree-creepers and bowerbirds. None of these branches has many species, and all are exclusively Australasian. This abundance of early types, along with the discovery in Australia of the oldest songbird fossils in the world, provides convincing evidence for the point of origin of songbirds: Australia.

Australia has repeatedly acted as a fountainhead for songbirds that have colonised Eurasia. One of the more recent immigrant groups is the orioles, which arrived in Eurasia from Australia/New Guinea around seven million years ago. Australian ecologist Ian Low thinks that Australia's songbirds are so successful because they exploited a new ecological niche that developed courtesy of Australia's infertile soils, which cause Australian plants to tend to hoard what nutrients they can get.[5] Nectar requires few nutrients to produce, and Australia's eucalypts produce abundant nectar in simple flowers. Visitors to Australia can easily see the consequences: flowering gum trees pulsate with lorikeets and the raucous cries of half a dozen species of honeyeaters. Relatively small species like songbirds have triumphed in this melee by becoming highly social, aggressive, and intelligent. If this is correct, the songbirds do not so much disprove Darwin's ideas about migration—which after all is premised on the idea that elevated competition drives evolution faster—but reveal an unexpected dimension of it.

Many mysteries remain concerning Europe's first 60 million years, but none is as vexing as the origin of one of its most extraordinary survivors. Known to Slovenes as the 'human fish' (and to the rest of the world as the olm), this blind, pink salamander grows to about 30 centimetres in length and is the only European vertebrate that spends its entire life in caves. In 1689, Johann Weikhard von Valvasor announced its existence in his book *The Glory of the Duchy of Carniola*. The dutchy, long since incorporated into Slovenia, was a small region, but Valvasor felt strongly that the world did not know enough about it. *Glory* filled 15 volumes with 3532 large-format pages, 528 copperplate engravings, and 24 appendices. The work was meticulously researched and scientifically accurate by the standards of the day, and to produce it, Valvasor installed a copperplate printery in Bogenšperk castle where he lived. The *Glory* bankrupted him, and he was forced to sell his castle, printery and estates. The Bishop of Zagreb took mercy on the patriot, purchasing his library and graphics for a handsome sum. But it was not enough and in 1693 Valvasor died a broken man, having survived the publication of his *Glory* by just four years.

There is something essentially European about Valvasor's grand obsession. After graduating from the Jesuit school in Ljubljana, he spent 14 years travelling Europe and North Africa, seeking the company of learned men. Edmund Halley proposed him as a member of the Royal Society, and in 1687 he became a Fellow. His *amor patriae* is evident in his great work, which must take its place among those volumes that from the time of Herodotus have sought to explain the nature of Europe, or some part of it. Thanks to the likes of Valvasor, Europe is alone among all the continents in having such a deep and rich written record of its natural history, the cost of which has all too often been paid in lives as well as fortunes.[*]

Valvasor's account of the olm stated that, after heavy rains, the creatures were flushed to the surface from underground caverns. The local people, he said, believed them to be the offspring of a cave dragon. Valvasor himself, however, characterised the olm as 'a worm and vermin, of which there are many hereabouts'. In his *Origin*, Darwin was

---

[*] *The Glory of the Duchy of Carniola* was originally published in High German. It was never republished, and for centuries was all but forgotten. Between 2009 and 2012, it was finally translated into Slovene by a team put together by Tomaž Čeč.

moved by the olm to exclaim, 'I am only surprised that more wrecks of ancient life have not been preserved'. By the nineteenth century a sort of olm-keeping mania had developed in Europe. Thousands of the creatures were exported, and some released into caves in France, Belgium, Hungary, Germany, Italy and possibly England.

The animal occurs naturally in certain caves and waters in Slovenia, Croatia, Bosnia and Herzegovina. Each population is slightly different, and nobody can agree on how many species there are. In 1994 scientists announced that they had found a black olm, which possesses eyes, and is restricted to the underground waters in a small area around Dobličica in the White Carniola region of Slovenia. Just how the olm got to Europe is mysterious. The Proteidae family to which it belongs includes just six species, five of which are North American. Fossils are rare, the oldest, from North America and dating from the end of the age of dinosaurs. The oldest European fossils are about 23 million years old.[6]

One thing everyone agrees on is that olms are bizarre. For a start, their entire life is lived in the slow lane. A batch of 64 eggs that was laid in Postojna Cave, Slovenia, took four months to hatch, and the young take the same amount of time as a human—about fourteen years—to reach sexual maturity. Nobody knows how long olms live, but at least a century seems a fair bet. And they can be hardy. One olm survived in captivity for twelve years *without eating*.

We have not treated olms well. For over a century they have been horrendously over-collected, and even used by farmers as pig-food. Today, the survivors are threatened by metal poisoning from industrial waste.[7] What a way to treat a national treasure!

CHAPTER 15

The Marvellous Miocene

Extending from about 23 to about 5.3 million years ago, the Miocene
was named by Charles Lyell.* It means 'less new' in Greek, and Lyell
gave it that name for the mouth-stretchingly mundane reason that he
considered that fewer Miocene species survived into modern times
than those of more recent epochs. Because of its favourable climate
and diverse fauna and flora, the Miocene is arguably Europe's most
enchanting epoch. A growing land area, enhanced migration corridors,
and a favourable climate conspired to create an unprecedented diversity
of mammals, some of which would go on to become successful colonisers of Asia and Africa. No longer solely a destination for immigrants,
Europe's fauna was beginning to influence surrounding continents.

Evidence of life in the Miocene is not evenly distributed across
Europe. Greece is spectacularly rich in reptile fossils, and Spain,
France, Switzerland and Italy have exceptionally good records from
both marine and terrestrial environments. Switzerland has yielded
some of the best-preserved fossil insects, and Germany some of the

---

* No single, global climatic event defines the beginning or end of the Miocene; its
onset being defined by the extinction of a species of plankton, and its termination by
an abrupt shift in Earth's magnetic poles.

most informative fossil plants. The British Isles, in contrast, which were a mainstay as we examined the Eocene and Oligocene, have almost no fossils of Miocene land mammals and plants.

The global cooling trend that began about 54 million years ago continued through the Miocene, though there were some striking reversals. For example, between 21 and 14 million years ago, conditions became as warm as they had been in the preceding Oligocene. When I think of Europe during this warm phase I imagine a sort of Côte d'Azur on the Seine. In the warmer conditions the seas rose, so the region that is now Paris was much closer to the coast than it is today. As the warming reached its height, much low-lying land became inundated, re-creating an island archipelago reminiscent of (though far more connected than) the one that had existed towards the end of the age of dinosaurs.[1]

Overall, however, the trend was towards more land, and greater connectivity. A serious phase of mountain-building commenced during the Miocene, and as the Alps and other mountains grew, the land-mass of Europe convulsed, causing volcanoes to erupt across the south, doubtless triggering many earthquakes. Some mountains must have risen with (from a geological perspective) astonishing speed: analysis of isotopic ratios of water and oxygen have revealed that the highest peaks of the Swiss Alps had reached their current elevation by the middle Miocene, some 15 million years ago.[2] The driving force behind this unrest was the vice-like grip created by Africa as it pushed north. And it was not just the Alps that were rising. As the land buckled, whole new islands and mountain ranges, separated by wide basins, rose.[*]

---

[*] The height of the Alps can vary by up to 27 centimetres, depending upon the pull of the moon, a fact that only became widely known when the particle accelerator at CERN, near Geneva, was built, and which works with such precision that the lunar pull had to be factored into its workings.

Among the most notable of these new mountain ranges was the Baetic Cordillera, which originated as a mountainous island incorporating the region around present-day Cadiz in southern Spain, the Sierra Nevada, the Rock of Gibraltar and the Sierra Tramuntana of Majorca (where the Majorcan midwife toad found refuge). To the north, the Pyrenees received a major uplift, as did Italy's Apennines. And further east, arcs of mountains developed from Albania right through to Turkey.

Many of Europe's most famous volcanic regions originated in the Miocene, as a result of great slabs of crust being driven into the molten mantle, where the rock was melted to magma. One important volcanic arc extends from Tuscany (Monte Amiata) to Sicily, where Etna, among other volcanos including Stromboli, remain active enough today to be major tourist attractions. Other potentially dangerous volcanic fields are dormant, including a large area south of Rome, which last erupted about 25,000 years ago. A second major volcanic region exists in Greece, where Methana, Santorini and Nisyros are considered active. During the Miocene, volcanic activity was far more widespread, with major volcanic provinces in the south of France, for example, during the epoch's beginning and closing phases.

Another defining feature of the European Miocene is extensive migrations. Following *la grande coupure* east–west migrations were often relatively unimpeded by topography. But extensive migration corridors also opened between Africa and Europe, which were at times so ample that, about 12 million years ago, the faunas of Kenya and Germany became almost indistinguishable.

Europe's vegetation continued to evolve, though much of the continent continued to be dominated by a subtropical forest rich in members of the laurel family, known as the laurophyllus forests. If you want to experience a laurophyllus forest, the Macaronesian archipelagos of Madeira and the Canary Islands are worth visiting. Macaronesia is

Greek for 'islands of the fortunate', and indeed these are fortunate isles, because on them a slice of ancient Europe survives to this day. The forested zone that occurs halfway up the mountains of Gran Canaria is a fine example, being dominated by four members of the laurel family, including the Canary and Azores laurel, the stinkwood and a relative of the avocado, all of which are ancient types growing alongside members of the ebony and olive families.

Another ancient vegetable relic surviving in Macaronesia is the legendary dragon tree, whose sap, marketed as dragon's blood, was in past times much prized as a medicine, as incense and as a dye. The dragon tree is not part of the laurophyllus forests; it is from a drier habitat that was beginning to establish itself in parts of Europe during the Miocene. These habitats were most evident on the Iberian Peninsula where, by 15 million years ago, members of the thorn-rose genus (*Neurada*), esparto grass, honey mesquite and nitre bush (*Nitraria*) flourished in arid shrublands.[3]

Sadly, Macaronesia's laurophyllus forests are almost devoid of the animal life that thrived in Miocene Europe. This is because the Macaronesian islands originated as volcanoes sprouting from the ocean floor, or as pieces of ocean crust that were pushed above the surface of the sea. The ancestral laurophyllus forests must have arrived as seeds in the guts of birds, or by floating over the ocean. With one important exception, land-based creatures could not cross the sea—and had you been a Carthaginian seafarer under the command of Hanno the Navigator in the centuries before Rome razed your great city, you may have seen those exceptional creatures with your own eyes.

Imagine being one of the first people to set foot on fabled Macaronesia. It is about 500 BCE, and back then, as today, the great peak of Teneriffe was so lofty that it was often lost in the clouds. But unlike the dry, rocky lowlands of Tenerife today, the island you experience is a verdant paradise, its trees filled with birds, including the soon-to-be-famous canary, which show no fear as you approach. Just one kind of large land creature exists there. Upon entering the forest, you see a huge reptile. The Tenerife giant lizard, *Gallotia goliath,* was a one-metre-long herbivore, with powerful jaws.

Today, all we have to remind us of its existence are bones and a single mummified head and chest found in a lava cave. Lizards nearly as large, but belonging to other species, lived elsewhere in the Canaries, but one by one they were exterminated when humans colonised the islands bringing with them predators such as dogs and cats. For a century it was believed that the giant lizards were no more. But then, at the very end of the last millennium, a remnant population was rediscovered.

For over a century the La Gomera giant lizard (*Gallotia bravoana*) was thought to be extinct. But in 1999 Spanish biologist Juan Carlos Rando discovered six individuals clinging to life on two inaccessible cliffs on the island. They had found a precarious refuge from predators and had somehow managed to subsist for many generations after their relatives, who had abounded throughout the island, had succumbed. A decision was made to bring a few of these last survivors into captivity, and today, courtesy of a painstaking recovery program, there are about 90 La Gomera giant lizards in wild and captive colonies. Perhaps one day, with a little help from humans, their progeny will reclaim their island home.

Lizards are known for their ability to reach islands, often by floating on rafts of vegetation, so it is not surprising that the gallotias colonised Macaronesia. Members of Europe's most diverse lizard family, the lacertids, their relatives include the wall lizards that can be seen throughout temperate Europe. The Canary Islands are also home to six species of smaller gallotias, which are similar in size to wall lizards, and survive in good numbers. They are remarkable for being among the world's smallest herbivorous lizards.

Biologists have long pointed to the giant gallotias as an example of the propensity of small-bodied creatures, when isolated on islands, to become giants. But a chance fossil discovery made near Ulm in Germany proved this to be wrong. The near-complete skeleton of a 22-million-year-old ancestral, gigantic, carnivorous *Gallotia* shows that when they reached Macaronesia, the lizards must have turned vegetarian, and many became dwarfs.[4]

Just after Macaronesia's laurels reached their island refuge, Europe's ancient forests began to change. Some of the best evidence of what happened comes from Germany, where the silicified trunks of about 80 species of trees have been found preserved in what was once a great lagoon fringing the northern slopes of the rising Alps. Dating from about 17.5 to 14 million years ago, they reveal the presence of a subtropical forest whose composition was shifting rapidly.[5] In the oldest layers of the deposit, the most abundant type of silicified wood comes from a relative of the cannonball mangrove (*Xylocarpus granatum*). A member of the mahogany family, the cannonball mangrove is today found in the tropical, coastal regions of Africa, Asia and Australasia, extending far into the Pacific.

In swamps nearby, palms and a relative of the Chinese swamp cypress known as *Glyptostrobus europeaus*, thrived. Some botanists consider the European fossil swamp cypress and the living Asian tree (*Glyptostrobus pensilis*) to be identical: 'it is possible that the tree now on the verge of extinction in China is the Tertiary species unchanged'.[6] The Chinese swamp cypress is deciduous, and it occurs naturally only on riverbanks and in swamps. Decay resistant, and with scented wood, it has been logged into near extinction.

On firmer ground away from the shore in Miocene Germany, a mixed forest of ancient relatives of the beech and myrtle trees grew. Just what clothed the higher slopes of the Alps we cannot know for sure, because no fossils are preserved. But it's a fair bet that towards the highest peaks, three kilometres or more above the sea, an alpine or subalpine flora was becoming established. Today, 350 of Europe's 4500 alpine plant species, including beauties like the *saxifrage du Mercantour* and the alpine poppy, are found nowhere else, suggesting that they have experienced a long period of evolution in their alpine home.

By the time the next layer of driftwood was being deposited in the ancient German lagoon, just a million or two years later, the climate had cooled and become drier. In this layer, the silicified trunks indicate that relatives of the acacias and members of the laurel family dominated a very diverse forest that included dipterocarps, which are among the tallest trees in the forests of Borneo today. In a more recent layer, which was formed in an even cooler and drier climate, oaks and laurels predominate, while in the most recent sediments (dating to around 14 million years ago) locust trees (*Robinia*), and members of the acacia family similar to the Kalahari Christmas tree (*Dicrostachys*), dominate.

These changes in vegetation—from tropical evergreen to deciduous and dryland types indicate some of the complexity of the floristic shifts that occurred in Europe during the 18 million years of the Miocene. The story is continued in the exceptionally rich fossil floras from south-western Romania. Dating to about 13 million years ago, they reveal a vegetation broadly reminiscent of, but much richer than, that found in contemporary Europe. Mixed forests of oaks and pines grew along the shores of an ancient lake, interspersed with beeches, elms, maples, hornbeam and some members of the laurel family. On boggy ground the swamp cypress flourished, alongside willows and poplars. Overall, this mixed vegetation of pines, evergreen and deciduous species resembles the forests still growing in east Asia and eastern North America, but also includes many genera that continue to dominate Romanian forests. This type of fossil flora, which occurs in many late Miocene and Pliocene deposits in Europe, is known as the Arcto-Tertiary Geoflora.

One strange feature of the European forests at the time the German and Romanian fossils were being deposited is the sudden appearance of the ginkgo. Although known from the age of dinosaurs, it seems to have become extinct in Europe sometime around or soon after the asteroid impact, so its reappearance some 40 million years later is surprising. But conditions in Miocene Europe clearly suited it and, for a time, ginkgoes flourished there.[7] The European ginkgo was not precisely the same as the today's ginkgo, which occurs naturally only in a tiny area in the mountains of China, but it was very close. Europe's ginkgoes appear to have become extinct sometime before the onset of the ice ages some 2.6 million years ago, with some of the last records coming from Romania. Their recent return to Europe, as street and garden trees should be welcomed as the return of a native.

By the end of the Miocene about five million years ago, mountain-building, dramatic drops in temperature, and lowering sea levels had created a Europe that was topographically broadly similar to the Europe of today. The cooling had also caused the extinction of cold-sensitive species in the European flora, and it is likely that grasslands, arid shrublands and alpine floras had become well-established. Of particular importance for the history of our own species, a mixed woodland-savanna occurred extensively in southeastern Europe—in what is today Greece and Turkey.

CHAPTER 16

A Miocene Bestiary

At times during the Miocene, the European fauna was almost as rich and diverse as that found in East Africa today. Those creatures have left an abundant fossil record that is at times bewilderingly varied, as well as fast changing. Rhinos had arrived in Europe during the Eocene but were only moderately diverse until the early Miocene. Then, between 23 and 20 million years ago a mosaic of habitats emerged that supported up to six coexisting species—including the small (a mere half-tonne) *Pleuroceros*, which had two horns side by side at the end of its snout. This, however, was only the beginning of Europe's rhino proliferation: by 16 million years ago the number of European rhino species had increased to 15, partly through local evolution and partly through the arrival of immigrants from Asia. However, as in earlier times, no more than five or six species co-occurred.[1]

The chalicotheres were among the strangest mammals that ever lived. They were perissodactyls—relatives of horses, rhinos and tapirs—and if all you had seen of one was its head, you might have mistaken it for a very strange horse. But its body was gorilla-like, and its limbs bore huge, sharp claws. The mix of features is so bizarre that for decades palaeontologists failed to recognise that fossils of the different parts came from the one type of animal.

The chalicotheres arose in Asia about 46 million years ago and spread rapidly to North America, and then to Europe, their migration taking the long way around, via the Bering land bridge and De Geer Corridor.[2] During the Miocene, a veritable evolutionary explosion of chalicotheres occurred in Europe, with no less than five genera existing at one time. The oddest of all was *Anisodon*. A denizen of Europe's late Miocene, it stood about 1.5 metres high at the shoulder and weighed about 600 kilograms. Its horse-like head was perched atop a long, almost okapi-like neck, which craned out from a body carried on long, stout forelimbs and short hindlimbs, giving it a steeply sloping back like a gorilla. And like a gorilla, *Anisodon* walked on its knuckles, folding its digits inwards to protect its sharp claws. It fed on foliage and seeds, nuts and hard fruits, whose shells were so tough they wore down its teeth to an extraordinary degree.[3] It was, in an ecological sense, Europe's answer to the ground sloths of South America and the gorillas of Africa.

The chalicotheres became extinct in Europe several million years ago, but one lineage survived in the forests of south Asia until about 780,000 years ago, so it would have been familiar to *Homo erectus*. If I possessed the godlike powers to resurrect just one creature from nature's graveyard, it would be *Anisodon*, an animal so unfathomable to me that it seems to belong in a fairytale.

Giraffes are often thought of as African, but in fact their origins lie in Asia, from where they migrated into Europe and Africa.[4] One extinct group known as the sivatheres grew very large and had antler-like growths on their heads. They probably looked like giant, horned okapis. The sivatheres became extinct about two million years ago, but other giraffes, including the ancestors of the living giraffe species and the okapi, thrived. The most abundant European Miocene giraffes belonged to the genus *Palaeotragus*, which was thought to have become extinct about five million years ago. Until 2010 *Palaeotragus* was of little interest to anyone but fossil giraffe specialists, but in that year two palaeontologists, Graham Mitchell and John Skinner, announced that *Palaeotragus* was not extinct at all. Instead, they opined, it still survived in the mountain forests of central Africa—in the form of the okapi.[5]

This is rejected by many scientists; but the claim that a supposedly extinct European creature might survive in the jungles of central Africa is astonishing. The okapi, with its purplish, velour-like coat and white rump stripes, has to be Earth's most beautiful mammal. If it is indeed an ancient European, or even an ecological replacement for one, rewilding enthusiasts in a distant future Europe warmed by anthropogenic greenhouse gases may seek to introduce it there.

Mitchell and Skinner, incidentally, also had surprising news about the origin of the modern, long-necked giraffes. This group, they claimed, probably evolved in Europe about eight million years ago, before spreading to Asia (where they became extinct) and Africa. As original Europeans, perhaps the long-necked giraffes also will one day be considered as candidates for reintroduction around the Mediterranean rim.

Bovids include an enormous variety of ruminants (cud-chewing, cloven-hoofed mammals) from antelopes to sheep and cattle, and are one of the most diverse and successful groups of large mammals ever to exist. Their origins lie in the early Miocene, when they diverged from the ancestors of the deer and giraffes. The oldest known bovid, *Eotragus*, was a dog-sized forest dweller that evolved in Eurasia, its short, straight horn cores and other bones being found in 18-million-year-old sediments from China to France.[6] Shortly thereafter the bovines (which include the cattle and their relatives) originated somewhere in Eurasia.

The antelopes originated in Europe about 17–18 million years ago—the oldest fossils (*Pseudoeotragus*) coming from Austria and Spain. Their spread—around 14 million years ago—to Africa and Asia, marks a great European success story. The caprines (a group that includes the goats, sheep and ibex) originated about 11 million years ago in either Africa or Europe, the earliest fossils coming from Africa and Greece. With the spread of grasslands around 10 million years ago all these bovids—which are superbly adapted to extracting nutrients from this fibrous resource—diversified rapidly.

Elephants originated in Africa and arrived in Europe 17.5 million years ago, probably via Asia.[7] The first to reach Europe belonged to a now-extinct family known as gomphotheres—primitive, four-tusked creatures. They became extinct in most places about 2.7 million years ago, when other kinds of elephants emerged from Africa and became widespread. But some survived in South America until humans arrived about 13,000 years ago: In geological terms, we missed out on seeing gomphotheres by a whisker.

About 16.5 million years ago two other kinds of elephants—deinotheres and mastodons—arrived on Europe's shores. *Prodeinotherium* (a deinothere) was about the size of today's Asian elephant, but its trunk was similar in size and function to that of a tapir. It lacked upper tusks, instead having a pair of downwards-pointing tusks in the lower jaws which may have been used to strip bark from trees. Over the Miocene the European deinotheres became enormous, some reaching weights of 15 tonnes—making them among the largest land mammals ever to have existed. Mastodons looked like living elephants, but the cusps of their molars resembled breasts (the name means breast-tooth), at least in the imaginations of some nineteenth century savants. They became extinct in Eurasia about 2.7 million years ago, but survived until 13,000 years ago in North America.

The advanced deer—the species with multi-pronged antlers that are shed annually—were present in Eurasia by around 14 million years ago. One, known as *Dicrocerus*, would give rise to the two great lineages of antlered deer, the Capreolinae and the Cervinae. Among the capreolines can be counted the roe deer, the moose, the reindeer and most American deer species (with the notable exception of the elk). The Cervinae include the muntjacs, the red deer and American elk, the fallow deer and the extinct Irish elk, as well as many Asian species including Pere David's deer and the chital.

The cervines are arguably Europe's greatest mammalian success story: the earliest type, *Cervavitus*, first appeared about 10 million years ago—in Europe. By around three million years later it had spread to east Asia and the cervines were well on the way to becoming the most abundant large herbivores in much of Eurasia.[8] When it was realised that cervines were European in origin, researchers were astonished, one writing: 'Europe should be considered more as a Dead End [in terms of migration] than an area with a normal evolutionary diversification.'[9]

Horses of the genus *Hipparion* (three-toed horses) migrated from North America into Europe 11.1 million years ago—one of the few successful migrant species at this time. Their arrival marks the start of Europe's Vallesian age—a subdivision of the Miocene.[10] Their fossils are extremely abundant, providing an easy way to date fossil deposits. Broadly similar in appearance to modern horses, they were about half the weight and had two small, hoofed side toes on each foot. Horses had been confined to North America for tens of millions of years, but were now able to migrate because a cold spell led to an expansion of the Antarctic ice cap, and with so much water frozen at the poles, sea levels dropped by 140 metres, opening a grassy land bridge across the Bering Strait.[11] Nothing like horses had existed in Eurasia, and they quickly filled the vacant ecological niche.

The strange dog-bears and bear-dogs of the Oligocene lingered into the Miocene in Europe, as did those primitive sabre-toothed predators the nimravids. The lynx-sized *Felis attica*, the ancestor of all living cats, was stalking the forests of ancient Greece and other parts of Eurasia by 12 million years ago, and the sabre-toothed cats were on the rise.[12] Fossil deposits, exposed by mining at Cerro de los Batallones near Madrid, reveal details of their evolution.[13] The deposits date from between 11.6 million and nine million years ago and are in filled-in gullies or caves. Most of the bones are from carnivores, indicating that the cavities functioned as natural traps, which were baited with the smell of rotting carcasses. An unusual find from the site is the bones of an extinct kind of red panda.

Complete skulls of several early kinds of sabre-toothed cats were recovered from Batallones, including an early member of the *Smilodon* lineage and a primitive kind of scimitar-toothed cat. The *Smilodon* ancestor was only the size of a leopard, while the scimitar-toothed cat was already as big as a lion.[14] The *Smilodon* ancestor was short-legged and almost bulldog-like in shape. Males and females were similar in size, suggesting that they were solitary ambush predators. Scimitar cats had a sloping back like a spotted hyena and were probably excellent runners. Males were much larger than females and may have had territories that overlapped with those of several females, like today's tiger.

The largest sabre-tooths, which survived until 13,000 years ago in North America, could kill young elephants. It's not known exactly how the sabre-like upper canines were used; but they were often broken, suggesting violent struggles by the prey. Some researchers believe they were used to sever the arteries in the neck; others think the sabre-toothed cats disembowelled their victims. Their sabres must have made it difficult to fit large hunks of flesh into their mouths, and their large, pointed incisors may have been used to pluck lumps of meat from the carcass. They may also have had rasp-like tongues covered in spines, like those of lions, to lick muscles from bones.

Hyenas evolved in Eurasia during the Miocene from ferret-like ancestors. They diverged into two types—the heavy-bone crushers, and fast running, dog-like types. These dog-like hyenas were extremely abundant in Miocene Europe, their fossils outnumbering those of all other carnivores in some 15 million-year-old deposits. But by five to seven million years ago, a changing climate, and possibly competition from the first dogs to reach Europe, caused them to decline. Today the only dog-like hyena is Africa's termite-eating aardwolf. The bone-crushing hyenas became the main scavengers of Eurasia—a role that they continue to play in Africa and Asia today. Part of their success seems to have lain in a partnership with the sabre-toothed cats, for the two carnivore types flourished together. Sabre-tooth cats had no ability to break up bones, so presumably the hyenas fed on the skeletons after the sabre-tooths had their fill.

But where were the dogs? They were still a continent away—in North America, awaiting a suitable land bridge to cross into Eurasia. Seven to five million years ago, at the very end of the Miocene *Eucyon*, a jackal-sized member of the dog family, made that crossing and spread quickly.[15]. But shortly thereafter they became extinct, and it would take another immigration from North America, about four million years ago, to bring new and larger canid species to Eurasia: these dogs would endure.

Ostriches strode the plains of Eastern Europe from the Miocene to the early Pleistocene. They all, with the exception of a diminutive species from Moldova, belonged to a single type, *Struthio asiaticus*, which was very similar to the living ostrich, but heavier. It is a scientific conundrum that while the fossil bones look alike, three distinctive kinds of fossil ostrich eggs have been found. Cariamas—metre-high, ground-dwelling predators that today are restricted to grasslands in South America—strode the Miocene plains of France, and parrots found a home in Miocene Germany. Many of the other birds inhabiting Miocene Europe were much like those you might see in Europe today.

We think of giant tortoises as denizens of islands, but in the past enormous chelonians could be found on every continent, and they thrived in Miocene Europe. Miocene python bones have been found in Greece and Bavaria, while a quarry at Wallenreid in Switzerland provided the world's oldest venomous snake fang.[16] Studies of a slightly more recent fang found in southern Germany indicate that these teeth were used to inject venom the same way that venomous snakes use their fangs today.[17]

The choristerans were crocodile-like reptiles with long, narrow jaws used for catching fish. In appearance and behaviour, they probably resembled India's gharial, yet they were entirely unrelated to crocodiles. Despite scads of research, the choristerans' placement on the tree of life remains unclear. But they seem to have originated before the dinosaurs evolved. By the Miocene, they were living fossils, unique to Europe. When scientists found the bones of a primitive choristeran in 20-million-year-old deposits in France and the Czech Republic they were surprised, describing the creature as having a 'ghostly lineage' of 11 million years—for no choristeran fossils are known from between 31 and 20 million years ago. They named the fossil *Lazarussuchus*, because it appears to have risen from the dead.[18] We do not know how long Lazarus lived after his resurrection, but *Lazarussuchus'* time in the sun appears to have been brief, for it is the last we hear of this venerable reptilian lineage.

CHAPTER 17

Europe's Extraordinary Apes

Apes seem alien to Europe today, but for about 12 million years, during the Miocene, the continent played a crucial role in their evolution. The ape family (Homindae), includes the human lineage, orangs, gorillas, chimpanzees. Very recent discoveries have revealed that the first hominids, the first bipedal apes, and possibly the first gorillas, all evolved in Europe. This would not have surprised Charles Darwin, who more than 100 years ago speculated that apes 'nearly as large as a man…existed in Europe during the Upper Miocene; and since so remote a period the earth has certainly undergone many great revolutions, and there has been ample time for migration on the largest scale'.[1]

The last common ancestors of the Old World monkeys and apes were monkey-like creatures called pliopithecoids. They probably originated in Asia, but soon spread to Europe and Africa.[2] Pliopithecoids persisted in Europe, Asia and Africa long after the first true apes and Old World monkeys appeared, and the fossil of one was the specimen Darwin was referred to. It was discovered in the 1820s by workers at a mine in Eppelsheim, near Mainz in Germany. The thigh bone was found in deposits containing the remains of many long-extinct creatures, and was long and straight, with a small hip joint. Overall it was so human-like that some nineteenth century savants posited that it must have belonged to a little girl.

This remarkable discovery was studiously ignored by Georges Cuvier. One of his dictums—which has not stood the test of time— was '*l'homme fossile n'existe pas*' (fossils of humans do not exist).[3] A devout Lutheran, he rejected any notion of evolution, instead proposing a theory of catastrophes and re-creations, which was more consistent with the Bible. Only the last cycle of creation, Cuvier posited, involved people—hence the lack of human fossils in older rocks. Although Cuvier managed to ignore the inconvenient femur, towards the end of the nineteenth century it was studied and named *Paidopithex rhenanus*. Today, it is thought to be from a late-surviving pliopithecoid that lived about 10 million years ago.[4]

Genetic studies indicate that the last common ancestor of the apes and Old World monkeys lived about 30 million years ago. But the oldest fossils, from Tanzania, are only 25.2 million years old.[5] The oldest ape, *Rukwapithecus*, is known from one partial jawbone, while the earliest Old World monkey, *Nsungwepithecus*, is known from a jaw fragment retaining a single molar. Scientists estimate that *Rukwapithecus* weighed about 12 kilograms, and that *Nsungwepithecus* was a little lighter. Beyond that, almost nothing is known of these important ancestors. The ape superfamily that *Rukwapithecus* gave rise to is known as the Hominoidea (it includes the gibbons as well as the orangs, gorillas, chimps and humans). Hominoids differ from monkeys in several ways, the most striking being the lack of an external tail. Apes do, however, retain tail bones, which have evolved into an entirely internal, curved structure known as a coccyx. Because apes and Old World monkeys share many similarities, the identification of a fossil tooth or limb bone as that of an ape is necessarily speculative. But a fossil coccyx constitutes gold-plated evidence.

For millions of years Africa had been bulldozing its way north. We often talk of continental drift, but the term is too passive: continents buckle, lift or smash whatever is in their path. About 19 million years ago Africa began to rotate counterclockwise to pinch the Tethys Sea in the region of what is now the Arabian Peninsula. A day must have come when sand touched sand. The great Tethys was severed, and in its place a land bridge linked Africa with the Turkish section of Europe. Elephants may have swum the narrowing Tethys Sea ahead of the connection, but apes abhor Neptune's realm: they would wait until a foot could be planted in dry sand—or perhaps even until a passage through a forest canopy could be negotiated.

The fossil ape *Ekembo* (previously included in *Proconsul*), from Kenya, lived between 19.5 and 17 million years ago. It was generally monkey-like, but probably lacked an external tail.* By around 17 million years ago, *Ekembo*-like apes had colonised Europe and undergone a rapid phase of evolution, transforming into griphopiths. Griphopiths are the earliest hominids, and their appearance in Europe at least a million years earlier than in Africa suggests that our family most likely arose in Europe—and not, as has been long assumed, in Africa. By 16.5 million years ago the Tethys seaway had again opened, isolating Europe's griphopiths. They continued evolving in isolation until about 15 million years ago, when the road to Africa again opened, allowing them to enter Africa and become established there.[6]

The 15-million-year-old African *Equatorius* was a recent emigrant, being very similar to the European griphopiths, though more terrestrial. Its contemporaneous relative *Nacholapithecus* (also African) provides the earliest unequivocal evidence of that key ape characteristic—the coccyx.[7] Beginning about 13 million years ago, the fossil

---

*Frustratingly, we cannot be absolutely certain of this based on the remains we have.

record of apes dwindles in Africa, before disappearing 11 million years ago. There are abundant fossils of all sorts of other creatures, so it looks as if the apes became extinct in Africa, perhaps through competition with Old World monkeys.

It may seem paradoxical that monkeys could outcompete apes. But if we absent humans and ask ourselves whether the apes or the Old World monkeys have done better in evolutionary terms, the answer is clear. There are about 140 living species of Old World monkeys, distributed from the icy mountains of Japan to Bali and from the Cape of Good Hope to Gibraltar. The apes, in contrast, comprise just 25 species which, with the exception of our own, are mostly rare inhabitants of African and Asian rainforests. Indeed, the monkeys have been displacing the apes from various habitats for many millions of years, so that today the surviving apes are mostly large species that have avoided competition with the highly efficient monkeys by increasing their body size.

It seems highly likely that about 13 million years ago, just before the great dwindling of the African apes, *Nacholapithecus*, or a species very like it, used another short-lived land bridge to cross from Africa into Europe. Some of the migrants didn't stay in Europe, however, but pushed on to Asia, where between 10 and 13 million years ago they gave rise to an ancestral orang utan. The apes that remained in Europe thrived, for their main competitors, the Old World monkeys, did not reach Europe until about 11 million years ago and did not become widespread there until about seven million years ago. Perhaps the more seasonal environment of Europe disadvantaged them.

CHAPTER 18

The First Upright Apes

There is little evidence of mammal migrations between Europe and Asia, and even less with Africa, between 13 and 10 million years ago. During this period, momentous changes began to occur among Europe's apes.[1] The story of that transformation is best told through the bones of an ancient Catalan, a Hungarian and a Greek. About 10 million years ago, in a waterway that is now a rubbish dump at Can Llobateres, near the town of Sabadell in Catalonia, the bones of creatures including ancient rhinos, flying squirrels, horses and antelopes began accumulating. In the summer of 1991, palaeoanthropologists David Begun and Salvador Moyà-Solà began to search there for fossils.[2] Ignoring the stench, they drove their picks into the sediment almost simultaneously, and to their astonishment exposed the skull of an ancient ape.

Over several years the partial skeleton of an extraordinary creature named *Hispanopithecus crusafonti* (Crusafonti's Hispanic ape) emerged from the clay. The bones make up the most complete hominid skeleton ever discovered in Europe.* Its limb bones reveal that *Hispanopithecus* moved about like chimps and gorillas. But a surprise came when

---

* The creature was named in honour of the Catalan palaeontologist Miquel Crusafonti i Pairó, who spent a lifetime studying the Miocene mammals of Iberia.

scientists examined its sinus cavities, which are large and of a shape seen only in gorillas, chimpanzees and humans. To judge from its sinuses, *Hispanopithecus crusafonti* is the earliest known hominine (the group including all great apes except orangs).

The bones of the second important specimen were unearthed in an iron mine near the town of Rudabánya in Hungary. The sediments exposed there were laid down in and around Lake Pannon, a now-vanished body of water that between 10 and 9.7 million years ago was about the size of the Great Lakes of North America. Freakish conditions at Rudabánya resulted in the capture of a 'snapshot' of an entire ecosystem.

Let's once more enter our time machine and visit the wonder that was Hungary 10 million years ago. We arrive at dusk in a wet and verdant world. The first thing we notice is the din of the evening chorus. The cries of ducks, pheasants, ravens, frogs and insects fill the air, and the early flying bats are already flitting about. The place feels more like modern-day Louisiana than central Europe.

A shower—one of many in this place that receives at least 1.2 metres of rain per year—has left the ground marshy. As we step away from the time machine we disturb a large beast. It's a tapir, emerging from the water and following the track left by a huge elephant—a deinothere—whose lower tusks have stripped bark from the trees beside the pad way. A mixed group of chalicotheres, rhinos and horses feed in woodland in the distance, stalked by a sabre-toothed nimravid and a hyena. The diversity of mammals is astonishing, with more than 70 present, including shrews, moles, bats, pliopithecoids, hares, many rodents including anomalures (strange squirrel-like creatures with scales on their tails, that can still be seen in central Africa), beavers and a wide variety of carnivores.[3]

Attracted by croaking we bend down and look into the reeds growing beside a small puddle, where we see two kinds of toads.

Picking up one of the smaller ones, we turn it over to reveal the rich yellow-and-black patterning of a firebelly. As we do so, the creature goes into its characteristic defensive posture, pushing its splayed legs high above its snout, creating a remarkable illusion that the toad's arse is in fact its head.

The larger kind of toad is gigantic. It bears little resemblance to its only living relative, the Hula painted frog of Israel. Calling loudly from among the reeds, it looks safe enough, but climate change will one day expel its entire genus from Europe. The firebellies and the painted frogs are members of the family Alytidae, which includes the midwife toads. The Miocene has been kind to these venerable creatures.

Something springs from the grass beside our feet and grabs a firebelly. It's a cobra, which, when it sees us, rises up and erects its hood. Soon cobras will be extinct in Europe, but this animal is entirely at home in the subtropical environment. There is one more surprise in store for us: a chimp-like call draws our attention and in the canopy we spy an extraordinary ape. Known as *Rudapithecus*, it was similar in some ways to *Hispanopithecus*, and is of exceptional importance to the story of human evolution.

The discovery of a skull of *Rudapithecus* at Rudabánya has greatly added to our understanding of this vanished ape. The specimen was found by Gábor Hernyák, a local geologist who had been working in the Rudabánya mine collecting fossils since the 1960s and had recovered many invaluable specimens. Hernyák had volunteered to work with David Begun on his 1999 dig at the mine but, according to Begun, had 'little patience with the niceties of documentation' of fossils, and without documentation the fossils are of reduced scientific value. So Hernyák was sent off to sweep up some loose dirt on a rock bench where the palaeoanthropologists had sat to eat their lunch. From millimetres below the surface that had supported their academic buttocks,

Hernyák uncovered the jaw of *Rudapithecus*, which, with more excavation, led to the discovery of the all-important skull.[4]

Rudapithecus was the size of a chimpanzee and had a chimp-sized
brain; older apes had much smaller brains relative to their body size,
and *Rudapithecus* provides the oldest evidence for such a large-brained
ape found anywhere in the world. Because of constraints on head size
during birth, large-brained apes are born while their brains are still
growing. In humans, this leads to a phenomenon known as the 'fourth
trimester'—the three months after birth when the brain is developing
fast but is now exposed to social stimulation. Some researchers think
that this phenomenon is responsible for our sociality and intelligence.[5]
If that is the case, then perhaps the foundations for these aspects of our
species began on the shores of Lake Pannon, some 10 million years ago.

About half a million years after a few individuals of *Rudapithecus*
died beside the great Pannonian Lake in what is now Hungary, a much
larger hominid was haunting the vicinity of what is now Athens and
Thessaloniki. *Ouranopithecus* was gorilla-sized, with heavy brow ridges,
big jaws and a palate that were all distinctly gorilla-like. Its molars,
however, were unlike those of gorillas but very human-like in that they
were covered with thick enamel. Its canines were also short, again like
our own and unlike the long, sharp canines of gorillas. *Ouranopithecus*
is as tantalising as it is frustrating, for this critically important link in
the story of hominid evolution has left us nothing but a few teeth and
jawbones, and a partial skull. We cannot know how it got around,
how large its brain was, or whether it had large sinuses. When it was
first discovered researchers described it as a possible ancestor of the
australopithecines—and therefore close to the human lineage. More
recently it has been suggested that *Ouranopithecus* is equally distantly
related to all African apes.

The discovery of some eight-million-year-old teeth from Ethiopia
has prompted yet another theory. They are claimed as the earliest

gorilla fossils, but they look very much like the teeth of *Ouranopithecus*.*
So it may be that *Ouranopithecus* is an ancestral gorilla, and that gorillas
evolved in Greece. If this is correct, then evolution has reversed itself
in several ways: first, the thin enamel of gorilla and chimp molars
must have re-evolved from the thick enamel of human-like molars;
and, secondly, the formidable canines of the gorillas and chimps have
evolved from short, human-like ones. If this is the case, the ancestor
of humans, gorillas and chimps had short canines and thick molar
enamel—features retained only in humans among living apes.

The importance of Greece to hominid evolution was given an enormous emphasis in May 2017, with the re-analysis of *Graecopithecus
freybergi*.[6] The story of this ape goes back to 1944, when threatened
German troops stationed near Athens were digging a bomb shelter. As
the soldiers dug desperately in the fine, reddish sediments, one exposed
the badly degraded jawbone of a primate. Just how the eroded bone,
which lacked tooth crowns, was noticed at all in the dire circumstances
and how it was preserved, are not recorded. Nor is there any hope of
re-excavating the site at Pyrgos Vasilissis, for the owners of the land, in
what is now virtually a suburb of Athens, have built a swimming pool
in the remains of the bomb shelter. Fortunately, the fossil can be dated
accurately—to around 7,175,000 years ago.

After the war, the find fell into the hands of the Dutch palaeoanthro-
pologist Gustav Heinrich Ralph von Koenigswald, who in 1972 named it
Graceopithecus freybergi—Freyberg's Greek ape.** Von Koenigswald was
famous for his researches on Java Man (*Homo erectus*) but he was risking
his hand in naming the miserable scrap of jawbone *Graecopithecus*.
Indeed, the name was widely considered a *nomen dubium*—a doubtful

---

* The *Ouranopithecus*-like creature has been named *Nakalipithecus*, and is known
from a jawbone and 11 isolated teeth.

** The specimen was housed in the Freyberg Museum.

name—and was in danger of being rejected by the International Commission on Zoological Nomenclature, which is a black mark indeed on the record of any zoologist. And there matters lay—until new technology revealed that the great professor had been right all along.

The roots of the premolars, it transpires, are key indicators of the hominin lineage, and CT scans of the tooth roots, along with the roots of a premolar found in Bulgaria, allowed the remains to be identified with some certainty as the oldest known hominin—that is, a direct ancestor of the upright apes including ourselves. This means that, in addition to democracy and gorillas, we must now credit Greece with being the cradle of the hominins—of which we humans are the only living representatives.

The reddish sediments that entombed the jaw have their own story to tell. Analysis of salt and minute rock particles show that they were carried to the Athens area from the Sahara, in dust clouds at least ten times larger than any seen today, indicating that the Saharan desert was already drying out seven million years ago and that its dust was falling on Europe in abundance. Elsewhere in the region, similar sediments have yielded the remains of ancient rhinos, horses, giraffes and large antelopes. Pollen from these sites reveals the presence of pines and oaks, saltbushes, daisies and grasses, while charcoal testifies to the occurrence of fire.[7] All in all, the dry, open environment that *Graecopithecus* inhabited was very different from the moist habitats favoured by Europe's earlier apes.

In 2017 an astonishing find was made near the village of Trachilos, on the island of Crete. There, sometime between 8.5 and 5.6 million years ago (with a most likely date of 5.7 million years ago), a pair of bipedal apes, perhaps accompanied by others, walked through the shallows on the edge of the sea, leaving tracks that would be preserved in great detail. At the time they were made, Crete was most likely a peninsula of mainland Europe.

The footprints left by these creatures vary in length between 9.4 and 22.3 centimetres, making them smaller than adult human footprints, but about the right size for *Graecopithecus*. They clearly show that the feet that made them had a 'ball' and a great toe aligned like ours. Only upright, walking apes have feet like this; it is probable that they were left by a relative of *Graecopithecus*, if not *Graecopithecus* itself.[8]

These footprints are the most recent evidence we have of hominins in Europe prior to the arrival of *Homo erectus* about two million years ago. It is touching to think that Europe's upright apes may not have survived for long after the Trachilos footprints were made, for towards the end of the Miocene Europe lost several species that continued to survive in Africa, including primitive giraffes like the okapi. The extinctions may have been caused by the same event that allowed the upright apes to migrate into Africa—the Messinian salinity crisis, when the entire Mediterranean dried out and a wide path to Africa was opened; though perhaps only briefly, before the basin became inhospitable to life.

The first possible hominin to appear in the African fossil record is *Sahelanthropus tchadensis*, which inhabited what is now Chad about seven million years ago. The next oldest is the 6.1–5.7 million-year-old *Orrorin tuguensis*, from Kenya. Known from a partial skeleton, it was definitely bipedal. Thereafter, Africa has yielded a rich array of upright apes spanning the gap between *Orrorin* and *Homo*. Mysteriously, almost no chimp fossils are known; a handful of half-a-million-year-old teeth from Ethiopia being the only ones identified to date.

Charles Darwin was right. Sometime about 5.7 million years ago, 'a migration on the largest scale' was made by apes walking from Europe to Africa. I'm sure that even the great man himself would have been surprised by the idea that the migration was made on two legs, rather than four. But following that event, until *Homo erectus* colonised Europe and Asia around 1.8 million years ago, the human story is all African.

A SUMMARY OF APE EVOLUTION DURING THE OLIGOCENE–MIOCENE

More than 30 million years ago	The ancestors of Old World monkeys and apes, the pliopithecoids, evolve in Asia.
25–30 million years ago	*Rukwapithecus* (ancestor of gibbons, orangs, gorillas, chimps and humans) evolves in Africa.
17 million years ago	Gryphopiths (ancestors of orangs, gorillas, chimps and humans) evolve in Europe.
13 million years ago	*Nacholapithecus* (last ancestor of orangs, gorillas, chimps and humans) evolves in Africa.
11 million years ago	*Hispanopithecus* (ancestor of gorillas, chimps and humans) evolves in Europe.
7 million years ago	*Graecopithecus*, the earliest ancestor of the human lineage, evolves in Europe.
6 million years ago	*Orrorin*, our direct ancestor, evolves in Africa.

CHAPTER 19

Lakes and Islands

Between about 11 and nine million years ago, mass migrations were transforming the faunas of Europe's fresh waters. The best place to see what happened is in sediments preserved around the ancient lakes of eastern and central Europe, including Lake Pannon. These extensive fresh waters allowed many new kinds of fish to colonise Europe, almost all of which came from Asia, which led to today's exceptionally rich faunas of the Danube catchment.[1]

Europe has about 600 species of freshwater fish, and 50 per cent of them belong to a single family, the Cyprinidae, which includes the carp, tench and minnows among others. Most of Europe's old, endemic species of freshwater fish are found in southern Europe, the northern European fauna having been destroyed by advancing ice, only to be re-colonised from the south following each ice age.

A remarkable survivor can be found in the southern Carpathians of Romania. *Romanichthys*, commonly known as the Romanian darter, is a very primitive cyprinid which has two dorsal fins and a covering of rough scales. Its discovery in 1957, in the upper reaches of the Varges River, caused ripples of surprise in the world of ichthyology. The development of hydroelectric dams has since had a severe impact on *Romanichthys*. It may survive in a single tributary of the Varges, but, without help, time is running out for this ancient Romanian.

Today, European waters are home to eight of the world's 27 species of sturgeons. They are an ancient breed of fish with a history going back more than 200 million years. Their fossil record, however, is so elusive that it's unclear just when they arrived in European waters. But they are adapted to life in lakes, and today the greatest diversity of sturgeon species occurs in the Caspian Sea on Europe's eastern border, where six species coexist. It's fair to assume that the ancestors of Europe's species arrived via Lake Pannon.

The beluga fish (not the whale) is the largest sturgeon, in times past reportedly reaching lengths of 5.5 metres and weights of 2000 kilograms in the Caspian Sea, making it one of the largest fish on Earth.[2] All sturgeon species are long-lived, some surviving for more than a century and taking 20 years to reach sexual maturity. They are, in effect, megafauna, and as for all of Europe's megafauna, they have fared badly on an ever more intensely populated continent. Illegal fishing continues to pillage the only viable sturgeon population left in the EU—in the lower reaches of the Danube in Serbia and Romania.

It is now time to turn to Europe's islands, as well as to one of its last, and perhaps most extraordinary, apes. So, let us enter our time machine and set the dials for the Mediterranean Sea about nine million years ago. Below us, the wine-dark waters are wide, but there is no sign of the Italian peninsula. Instead, two great islands are visible, parts of which will in time be incorporated into mainland Italy. Both have left a rich fossil record.

We land on the lost island of Gargano, a place that existed between 12 and four million years ago, and step out into the balmy air. Before us is a gullied limestone plateau, covered in a mixed vegetation of forest and more open habitats. A shadow passes over us. We look up and see a falcon the size of an eagle swooping to investigate. It disturbs a group of *Hoplitomeryx*. Rather goat-like in size and shape, they have five horns on their head, one of which springs from between their eyes, giving them a fierce look that is accentuated by the long, sabre-like upper canines. Despite appearances, they are herbivores—a kind of horned deer—and the largest of Gargano's inhabitants. The remains of five species have been discovered (they may have existed at different times), the largest the size of a red deer.

The startled *Hoplitomeryx* canter towards a thicket, and an ugly, pig-eyed creature that looks to be all head, rushes out and grabs a fawn, snarling and struggling to subdue its prey. *Deinogaleryx* is the largest hedgehog ever to exist. One-third of its 60-centimetre length is head, the rest a hairy body with short legs. Its incisors stick out almost horizontally from its ferocious maw, while its tiny eyes give it a particularly vicious aspect. In the absence of cats and other carnivores, evolution has recruited this most unlikely creature to be Gargano's top mammalian carnivore. But the titanic hedgehog was not the only predator of ancient Gargano. Had we time to explore further we might see a gigantic barn owl, which at over a metre tall was twice the size of the largest owl living today. Add a giant, flightless goose, an endemic otter, a giant pika (a rabbit-like creature), five species of dormice, some of which were giants, and three gigantic hamsters, and you have a very strange fauna indeed.

The bones of Gargano's ancient inhabitants were preserved when the island's limestone plateau eroded into a cavernous formation that trapped and preserved them. Most, if not all, of the island was then submerged and was overlain with a blanket of marine sediments. As the boot-shaped peninsula of Italy took shape, it gave a backwards kick, so to speak, rotating from a position adjacent to Sardinia to one nearer to the eastern Adriatic coast, colliding with the then submerged island of Gargano, and elevating it to about 1000 metres above the sea before fusing it to the Italian peninsula to become the 'spur' on the boot.

We return to our time machine and travel west, to Tuscania, the largest island of Miocene Europe. Composed of what today are the islands of Sardinia and Corsica, as well as parts of Tuscany, Tuscania was larger than any modern Mediterranean island. Over the last 50 million years it has been intermittently connected to the European mainland, allowing new species to colonise. By about nine million years ago, however, a prolonged period of isolation led to the development of a most unusual island fauna. Our time machine touches down beside the estuary of a tropical river, on a high dune separating a broad stretch of swamp forest from the sea.

As we step out, herds of small-to-tiny antelopes, clearly belonging to two distinct species, accompanied by a much larger primitive giraffe, browse on the sparse vegetation of the dune.* The larger of the antelope species is the most abundant herbivore on the island, and has distinctive spiral horns. The smaller one, barely the size of a hare, has simpler, curved horns. The giraffe (whose fossils are few) may have resembled a small okapi. In the shallows a dwarf, buffalo-like creature stands, cooling down, accompanied by an Etruscan pig—a small, short-snouted porker.

An unusual ape ambles onto the dune. The gibbon-sized creature is walking upright with an awkward gait, holding in its right hand a broad leaf to protect its head from the sun. It walks towards a clump of mangroves and climbs into the canopy where it feeds on the salty leaves. The Tuscanian ape, *Oreopithecus bambolii*, is by far the best known of all of Europe's apes, for entire skeletons were discovered in a lignite mine in Tuscany. They reveal a creature weighing between 30 and 35 kilograms, with long arms, a small, globular cranium and teeth adapted to eating leaves. It was not intelligent, its brain being only half the size of the brains of other early apes.

---

* The precise identity of the 'giraffe', *Umbriotherium azzarolli*, is still disputed, but certain features of its premolars resemble those of primitive giraffes.

As its long arms and leafy diet indicate, *Oreopithecus* was primarily adapted for life in the treetops, moving through the canopy like a gibbon, swinging arm over arm. But that is not the whole story. Its spine is curved in a very distinctive way, and its pelvis is astonishingly human-like, suggesting that it habitually stood upright. Moreover, each foot has a great toe that sticks out at a 90-degree angle, providing a sturdy tripod on which to balance. *Oreopithecus* is a mystery hiding in full view: we could hardly want for more skeletal evidence, yet scientists can't agree on where it lies on our family tree. Was it an upright hominin—and thus on the human lineage—or a more primitive type of ape that independently evolved the ability to stand on two legs?

Oreopithecus was one of Europe's last apes. Had we arrived on Tuscania around six million years ago and looked northwards, we would have seen a distant shore across the sea. For generation after generation, that shore would have moved imperceptibly closer, bearing its freight of hyenas, sabre-tooths and primitive canids that lurked in the forests behind the beaches of mainland Europe. When shore finally touched shore, the little ape would not have stood a chance.

*

If you have ever visited Monaco, to play at Monte Carlo, perhaps, or watch the grand prix, you may almost have rubbed shoulders with an intriguing American. Not Princess Grace, but Strinati's cave-salamander—which deserves to be every bit as celebrated and treasured as any actor or head of state. Just 10 centimetres long, of a retiring nature and—strange for a terrestrial organism—lacking lungs, it gets by by breathing through its skin. The skin must be kept moist, which is why it spends most of its life in caves, crevices and other humid places, emerging only at night to feed, using its long, projectile tongue to catch insects and other small creatures, much in the manner of a toad.

The origins of this retiring creature have kept scientists guessing for more than a century. When did its ancestors arrive in the limestone fastness of Monaco, and how did they get there? Strinati's cave salamander is one of just seven European cave salamanders, four of which are found only on the island of Sardinia, the others being distributed in southwestern France and Italy, San Marino and Monaco. One might say that their fondness for tiny nation states is almost as much of a mystery as their origins.

The family group Plethodontidae, to which the European cave salamanders belong, contains about 450 species, making it the largest of the salamander and newt families. And 98 per cent of its species are restricted to the Americas. All lack lungs, though that handicap seems to have counted for little. In Mark Twain National Forest in Missouri, for example, they are—if you count them by weight—the dominant form of life, with 1400 tonnes of plethodontids lurking in the leaf litter and wetlands of its 600,000 hectares.

Europe's cave salamanders, scientists agree, must have come from North America. But when, and by which route? Did they, like the amphisbaenids, arrive in the wake of the dinosaur extinction? And did they travel overland, or by sea? Some researchers suspect that they are ancient relics that have survived only by retreating to their subterranean fastnesses. Their distribution—which until recently was thought to include only the Americas and Europe—supported the idea that they must have crossed a land bridge between the two landmasses, perhaps during the age of dinosaurs. But the oldest fossils of the group in Europe, which come from Slovakia (where they no longer occur), date only to the middle Miocene—around 14 million years ago.[3]

In 2005, a remarkable discovery was announced. An American teacher working in Korea was leading his students on a walk in Chungcheongnam-do when he spotted a salamander in a rocky crevice. He captured the animal and sent it to Dr David Wake, an expert on salamander classification, who proclaimed it 'the most stunning discovery in the field of herpetology during my lifetime'.[4] It was a lungless salamander—the first ever found in Asia. The discovery makes it likely that lungless salamanders arrived in Europe, via Asia, during the Miocene.

CHAPTER 20

The Messinian Salinity Crisis

Since the nineteenth century, geologists have known that layers of salt and gypsum exist around the Mediterranean, but until 1961 nobody understood how they got there. In that year a seismic survey was conducted, revealing a layer of salt, in places more than one and a half kilometres thick, below the entire Mediterranean basin. Astonished scientists conducted a drilling program, and a decade later confirmed that the layers of salts and other evaporites could mean only one thing: at some point the Mediterranean Sea had dried out. A research program found that the great drying commenced about six million years ago during the Messinian age, the last stage of the Miocene.[*] Known as the Messinian salinity crisis, it resulted from the clockwise rotation of Africa, which closed the Strait of Gibraltar and isolated the Mediterranean from the Atlantic Ocean.

You might think that mighty rivers like the Rhône, Nile and Danube that flow into the Mediterranean would prevent the sea drying out, even if it were cut off from the Atlantic. But so great is the amount of water that evaporates each year from the Mediterranean

---

[*] The Messinian age is named for layers of evaporite rocks that outcrop near Messina, Sicily.

that all of the water brought in by rivers, along with all of the rainfall it receives directly, cannot offset it. In fact, the rivers draining into the Mediterranean bring in only about a tenth of the amount of water lost through evaporation. The rest of the water deficit is replaced by flows from the Atlantic, which is why a swift current flows through the Strait of Gibraltar. Without this Atlantic water, the level of the Mediterranean Sea would drop at a rate of a metre per year.

When the connection with the Atlantic was blocked, it took just 1000 years for the Mediterranean to dry out, creating a vast salt plain, more than 4000 metres below sea level at its lowest point, dotted with hypersaline lagoons. The Mediterranean's islands now towered as high as seven kilometres above the salty plain, where temperatures may have reached 80° Celsius—a phenomenon that must have profoundly affected regional atmospheric circulation and rainfall, and precluded all life except bacterial extremophiles.*

The drying of the Mediterranean caused the rivers flowing into the basin to cut deep valleys. For example, the Nile flowed 2.4 kilometres deeper than the level of Cairo, while the Rhône cascaded down a steep slope, creating a valley 900 metres deep below present-day Marseilles. The Mediterranean did not remain continuously dry during the Messsinian crisis: as the climate altered, it partially filled periodically, leaving a series of salty and less salty layers in the sediment. Around 5.3 million years ago, after some 600,000 years, a link with the Atlantic was re-established when rivers draining into the basin cut through the barrier.

Once the ocean water found a way into the basin, it cut a deeper channel, and thus began the so-called Langelian flood, in which the waters of the Mediterranean rose at rates of up to 10 metres a day.

---

* It is difficult to be more precise about the highest points of the Mediterranean Islands six million years ago.

Initially, the waters descended the four vertical kilometres to the salty basin floor through a series of cascades that followed a relatively gentle slope. Nonetheless, it must have been an awesome sight that overall would have dwarfed any waterfall existing today. Within a century, the Mediterranean was refilled.

The Messinian salinity crisis changed the world. The global sea level rose by 10 metres because the water evaporated from the Mediterranean was added to the seas elsewhere, and during the century it took for the Mediterranean to refill, the oceans fell by 10 metres. So much salt—about a million cubic kilometres—was locked up in the sedimentary layers underlying the Mediterranean that the salinity of all of Earth's oceans remains reduced. Because freshwater freezes at higher temperatures than salt water, the surface layers of the oceans near the poles froze more readily. As the climate continued to cool, this would hasten the onset of the ice ages.

The end of the Miocene is dated to 5.3 million years ago. Although it roughly coincides with the end of the Messinian salinity crisis, the Miocene's end is not defined by this event. Indeed, it is not marked by any global cataclysm, but rather by the extinction of an obscure and tiny plankton known as *Triquetrorhabdulus rugosus*. Geologists often choose the extinction of a species of plankton to define the end of a geological period, because the tiny fossils are widespread and easily found, making it possible for palaeontologists to trace the event globally.

This is sound science, but the poet in me chafes against it. Surely the beginning of a new geological epoch is portentous and should be marked by more than the passing of a microscopic algae? One such possibility for the dawn of the Pliocene is the origination of the genus *Gadus*, which is significant because it includes that most economically important fish, the cod.[1] Europeans have enjoyed 'fish and chips', *bacalao* and other cod-based delicacies for centuries, so surely a case can be made for this fish to be the herald of the Pliocene. Yet I sense I'm fighting a losing battle—to take a small liberty with Philippians 4:7: the ways of geologists can, like a piece of cod, passeth all understanding.

CHAPTER 21

The Pliocene—
Time of Laocoon

If we cannot define the advent of the Pliocene by the rise of cod, then perhaps we should abolish it altogether. It is after all ridiculously brief and has nothing much to distinguish it from the Miocene. As currently defined, it extends from just 5.3 million to 2.6 million years ago. Named by Charles Lyell, a rough translation is 'continuation of the recent'. The great man appears to have slipped up when minting the name—so egregiously indeed that the lexicographer Henry Watson Fowler, of *A Dictionary of Modern English Usage* fame, lambasted the name of the epoch as a 'regrettable barbarism'.* Lyell justified it on the rather lame basis that many Pliocene molluscs are similar to living species. But what is really characteristic of the Pliocene, in Europe at least, is that it was a time of giants. In effect, the Pliocene is Europe's last, great flowering, after which the continent's biodiversity entered a decline.

A map of Pliocene Europe has that uncanny quality of being familiar, yet not quite right. Looking to Iceland's east we see that the entirety of Scandinavia is not so much missing as combined into a massy lump of land that forms a northwestern bulwark of Europe.

---

*In Fowler's terms, a barbarism is a word minted using words from more than one language.

This is because the basin of the Baltic Sea is yet to be carved from the rock. And where is Britain? Like Scandinavia, it is embedded in a broad peninsula, this one projecting northwards from present-day France. Neither the English Channel nor the Irish Sea exist. To the south, the form of the Mediterranean lands is even more disconcerting. Beginning in the west, the Baetic Cordillera (comprising the Sierra Nevada and the Balearic Islands) is still a single, mountainous island located at the entry to the Mediterranean, where the Strait of Gibraltar is today. Tuscania lies to its east, attached to the mainland by a peduncle as if hanging from the Maritime Alps. Italy, meanwhile, is broadly connected to Turkey, mainland Greece is a lesser peninsula, while parts of eastern Europe as far north as Romania are under the waves.

How to account for these many differences? Sea levels were 25 metres higher in the early Pliocene than they are today. Yet many parts of Europe that are under the sea now were dry land then. This is because, in the north, the erosion caused by subsequent glaciers and frozen sheets of the ice age have chiselled away at the land, carving channels and gulfs that give northern Europe its current topography. But much of the work in shaping contemporary Europe's south was done by the restless energy of the tectonic plates, driven by a northward-moving Africa.

Average global temperatures during the Pliocene were 2–3° Celsius warmer than they are today, and until three million years ago the northern ice cap formed on the Arctic Sea only during the winter. But the climate was cooling, and Europe was becoming drier and more seasonal, favouring the spread of deciduous and coniferous forests across the north. Prior to the ice ages—right up to the end of the Pliocene— Europe's forests were broadly similar to those of North America and Asia today. They were made up of a great number of species, including wing-nut (a relative of the walnut), hickory, tulip-tree, hemlock, blackgum, sequoia, swamp cypress, magnolia and liquidambar, which are no longer found in Europe, along with familiar European types such as oak, hornbeam, beech, pine, spruce and fir.*

Botanists refer to this vegetation type as the Arcto-Tertiary Geoflora. Its loss from Europe at the end of the Pliocene is called the Asa Gray disjunction, after the great nineteenth century American botanist who so convincingly explained its causes. At the time Gray worked, the ice ages were a mystery, though it was clear that in the distant past the earth had been far colder than it is today. Gray argued that the most cold-sensitive trees of the Arcto-Tertiary Geoflora had been squeezed against the alps by the growing chill, until they were exterminated. Asia and North America, in contrast, have uninterrupted, forested coastlines that run from equator almost to the pole, providing a migration corridor for species as the climate changed.[1]

Asa Gray's concept reverberates through moral, philosophical and cultural dimensions of European landscapes. Without his work we would see the glorious golden autumn foliage of a liquidambar, or a spring magnolia in full bloom, as alien to Europe. But such trees are mere prodigals, albeit prodigals forced two million years ago from their native homes, and now returning, courtesy of colonial-era botanists and a warming climate.

Incidentally, over the millennia, Asia has provided a refuge for far more of Europe's biological heritage than just the Arcto-Tertiary Geoflora. Many organisms that became extinct over Europe's long history have survived in the rainforests of Malaysia and in regions to its north and east. For example, close relatives of the nipa palm and water cypress that grew in Germany 47 million years ago continue to thrive in Malaysia. The cannonball mangrove that thrived in Bavaria 18 million years ago can still be seen in the Indo-Malayan archipelago. And

---

* Liquidambar maintains a European foothold in a small area of southwestern Turkey.

remember the saratoga fish from Hainin, and the pig-nosed turtles from Messel? Europeans can effectively time-travel to the distant past of their continent by boarding a jet bound for the Malay Archipelago.

Some of the most intriguing creatures ever to inhabit Europe lived during the Pliocene, and the most fascinating of all have, tragically, been lost forever. The remains of one remarkable animal were recovered during what is arguably the last of Europe's religiously inspired wars—the Crimean campaign of 1853–56. During the conflict, as the naval and land assaults on Sevastapol dragged bitterly on and the Light Brigade carried out its fatal charge, Captain Thomas Abel Brimage Spratt, commander of HMV *Spitfire*, rendered distinguished military service, in recognition of which he was made a Most Honorable Military Member of the Order of the Bath. Spratt was a man after my own heart. Somehow, amidst the shell and rifle fire, he found time to look for fossils, and as he rummaged the rocks near Thessaloniki, he discovered something quite special. He returned to Britain with his collection, and in 1857 the great anatomist Sir Richard Owen set to work identifying the specimens Spratt had passed on to him.

Owen began his career at the Royal College of Surgeons. He was a horrible man; his biographer Deborah Cadbury saying of her subject that he 'had a penchant for sadism' and was 'driven by arrogance and jealousy'.[2] He was perhaps at his worst when dealing with his great rival in describing dinosaurs, Gideon Mantell. Mantell, who had discovered the dinosaur *Iguanodon*, a feat that Owen so envied that he claimed to have discovered the creature himself. As the rivalry between the pair escalated, Mantell said of Owen that it was 'a pity that a man so talented should be so dastardly and envious'. Over the years Mantell named four of the five dinosaur genera then known, which only acted to fuel Owen's envy.

Mantell was a medical doctor, but he was so absorbed in his palae-ontological research that his practice suffered. He moved to Brighton on England's south coast in the hope of better fortunes, but he was soon destitute and was forced to sell his fossil collection to the British Museum, where Owen was already influential.* Mantell asked for £5000 but settled for £4000—a poor price indeed for an arrangement that placed the fruits of his lifetime of palaeontological labour at his rival's disposal. But poor Mantel's degradation did not end there. In 1841 he suffered a carriage accident in which he fell from his seat and became entangled in the horses' reins. As he was dragged across the ground, his spine was gravely injured. To deal with the ongoing pain he took opium, but in 1852 it all became too much, and the good doctor overdosed. After his death, in a reptilian act, Owen had someone remove the damaged section of Mantell's spine, which he pickled and stored in a jar, and it joined Mantell's dinosaurs as one of Owen's trophies.

Owen rejected out of hand Darwin's theory of evolution, perhaps in part because he was as cunning a politician as he was a brilliant anatomist. Yet somehow his scientific reputation survived even his dogged adherence to Creationism. In fact, the terrible truth is that Sir Richard Owen, KCB, FRMS, FRS, president of the British Association for the Advancement of Science and darling of the nobility, got away with almost everything. For 90 years—until 2008—his statue held pride of place at the top of the grand staircase in the British Museum of Natural History. And Mantel's spine languished in its glass jar at the Royal College of Surgeons until 1969, when it was destroyed to free up space.

Owen fancied that he knew the internal structure of every creature on earth, but the fossils Spratt had collected near Thessaloniki forced him to expand his studies. Spratt's thirteen bones, Owen concluded, could belong to no other snake than a viper. What was confounding, however, was its size, for the bones must have come from a creature at least three metres long. To explain this, Owen resorted to the classics:

---

*Owen would take over the Natural History department there in 1856.

The classical myth embalmed in the verse of Virgil and embodied in the marble of the Laocoon, would indicate a familiarity in the minds of the ancient colonists of Greece with the idea at least of serpents as large...But according to actual knowledge, and any positive records of zoology, the serpent...must be deemed an extinct species.[3]

Owen named the remains of what was clearly a very large and formidable viper *Laophis crotaloides*—the 'rattlesnake-like snake of the people'.[4]

It strains credulity that such an important fossil as *Laophis* could be lost by the British Museum, but lost it was, and for almost 160 years the giant viper of Thessaloniki was all but forgotten. Then, in 2014, a group of researchers announced the discovery of a single partial snake vertebra—barely two centimetres across—at Megalo Emvolon near Thessaloniki in northern Greece. It was about four million years old, and it clearly matched drawings of the lost bones of Owen's near-mythical viper.

The sediments the bone was preserved in were formed in an ancient lake which, to judge from fossil pollen, was surrounded by sparsely wooded grassland. The fossil fauna discovered alongside the remains of the great snake is reminiscent of that found in the seasonally dry parts of northern India today, including extinct horses, pigs, giant tortoises, a species of monkey, rabbits and a giant peacock.[5] Fragmented as the vertebra was, the researchers concluded that *Laophis* was the largest viper that ever lived. The monster seems to have been closely related to the viper genus (*Vipera*) that inhabits Europe today, though the largest living *Vipera*—the horned viper of southern Europe and the Middle East is, at less than a metre long, just one-third its length.

The weight of snakes increases disproportionately with length, and the three-metre-long *Laophis* is estimated to have weighed

26 kilograms, which makes it more than two and a half times the weight of the king cobra, the largest venomous snake alive today.[6] What did this great viper feed on? Today's horned vipers feed on mammals (mostly rodents), birds and lizards; perhaps *Laophis* feasted on monkeys, rabbits and giant peacocks. All that we can say with any certainty is that at the dawn of the Pliocene, Europe was home to the greatest venomous snake that has ever existed.

The giant tortoises that shared *Laophis*'s habitat were also some of the largest that ever lived. *Titanochelon* was truly stupendous: with shells reaching two metres in length it was the size of a small car. Unique to Europe, these gargantuan chelonians looked like Galapagos tortoises, but were far larger. Giant tortoises need warm conditions, as they can't burrow as smaller tortoises can. As the ice ages set in, they were restricted to the south of Europe and, as so many other species did, they made their last stand in Spain. The most recent bones, found in an ancient hyena den on a floodplain, are about two million years old.[7] And with the giant chelonians went Europe's last crocodiles and alligators—all carried off by the increasing chill, though it seems possible that the arrival of *Homo erectus* from Africa might also have played a role in the extinction of the tortoises. After all, the fossil record speaks eloquently of the fact that upright apes and great tortoises don't mix.

With the cold and the spread of grasslands the bovids flourished. Only two of the nine tribes—the Bovini and Caprini—diversified greatly in Europe.[8] The Bovini, which include cattle, bison and buffalo, appear in the European fossil record in the early Pliocene and quickly proliferates. The Caprini, which include goats, sheep and ibex, also diversified during the Pliocene.

Throughout the period, toothy giants persisted in the oceans. Perhaps the most spectacular was the megalodon shark. The largest predator in Earth's history, it reached lengths of 18 metres and a weight of 70 tonnes. It was named by the Swiss naturalist Louis Agassiz in

1835, who studied some of its enormous teeth, the largest of which are 18 centimetres long and weigh more than a kilogram. The beast had hundreds of them in its jaws and, as befits such a monster, it ate whales. Megalodon's bite force was between five and ten times greater than that of the great white shark. Gouges are common on the flipper and tail bones of fossil whales, suggesting that megalodon would bite off the locomotor appendages before feeding on the disabled beast. Megalodon evolved early in the Miocene and just kept getting bigger. The very largest individuals lived in the Pliocene—just prior to the species' extinction about 2.6 million years ago.[9]

On land, more giants found their way into Europe. After a hiatus of more than 10 million years, elephant migrations recommenced, bringing new species into Europe, while the descendants of earlier migrants declined to extinction. The ancestors of all living African and Asian elephants, as well as the mammoths, originated in Africa towards the end of the Miocene. Mammoths migrated to Europe from Africa around three million years ago and soon gave rise to *Mammuthus meridionalis*, a species weighing 12 tonnes and adapted to life in Europe's woodlands.[10] A relative of the Indian elephant also arrived in Europe in the late Pliocene, but soon became extinct there, as did Europe's deinotheres and gompotheres.[11]

The Pliocene heralded the arrival of Europe's first modern bears. The Auvergne bear, *Ursus minimus*, was similar to, but a little smaller than, the Asian black bear. It appears to have given rise to the Etruscan bear (*Ursus etruscus*), which is so similar to the Asian black bear that some researchers consider them to be one and the same. In a fairytale-like twist, the Etruscan bear gave rise to Europe's three bears of yore: the brown bear, the cave bear and the polar bear.

I cannot leave the Pliocene without saying *vale* to those tiny and obscure amphibians, the pert'uns. After enduring almost 350 million years, they finally winked out 2.8 million years ago, the last we see of them being some bones preserved in limestone fissures near Verona. Had they survived we would marvel at them as being among the most venerable of Earth's creatures.

The composition of Europe's fauna at the time the ice ages set in is something of a mystery; fossil sites are few, and the possibilities of migration were many and varied.[12] A rich two-million-year-old fossil deposit in southern Spain offers a window into this 'lost world'. It has yielded the remains of 32 mammal species, including a primitive kind of muskox (clearly adapted to much warmer conditions than the kind living today) a wolf, a giraffe, the brown hyena and the bush pig, the last two of which are otherwise unknown in Europe, but thrive in Africa today. Analysis of the fossils has allowed Dr Alfonso Arribas and his colleagues to develop a simple hypothesis of migration and, according to Occam's razor, the simpler the explanation the more likely it is.*

Arribas and his team think that Europe's early ice age fauna resulted from just one migration event that occurred almost two million years ago, and which took place across islands in what is now the Strait of Gibraltar. Even the Asian species used this route, the researchers argue, after migrating right across north Africa. The theory was challenged just a year after it was articulated, when experts in the evolution of the dog family announced that they had detected the presence of the earliest wolf-like creatures, *Canis etruscus*, in French fossil deposits dating to around 3.1 million years ago.[13] We are, I think, still a long way from a full understanding of the migrations that occurred in Europe on the eve of the ice ages. And only more careful digging will provide the answers.

---

* William of Occam was an English Franciscan Friar who lived in the fourteenth century. He is remembered for his dictum that: 'among competing hypotheses, the one with the fewest assumptions should be selected'.

III

ICE AGES

2.6 Million–38,000 Years Ago

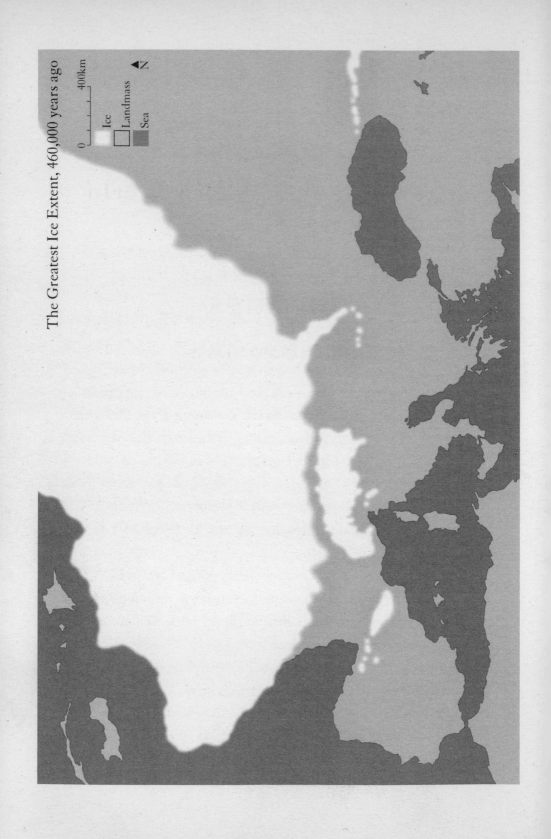

The Greatest Ice Extent, 460,000 years ago

Ice
Landmass
Sea

N

0   400km

CHAPTER 22

The Pleistocene–
Gateway to the Modern World

In 2009—the same year that Arribas and his colleagues published their stimulating research on the two-million-year-old 'late Pliocene' fauna of southwestern Europe—the panjandrums at the International Union of Geological Sciences moved the beginning of the Pleistocene back by more than half a million years—from 1.8 million years ago to 2.6 million years ago. Their reasoning was that the glacial cycles (of which the ice ages are a part) should be included in their entirety in the Pleistocene, and that the first glacial cycle began 2.6 million years ago. It was a worthy and sensible decision, not least because it further reduced the rump-like Pliocene.

It was that veteran namer of geological periods Charles Lyell who coined the term *Pleistocene*. It means 'most new', in recognition of the fact that about 70 per cent of mollusc fossils from the Sicilian deposits studied by the venerable professor belonged to still-existing types. While the beginning of the Pleistocene has been appropriately designated, I cannot say the same for its termination. The International Union of Geological Sciences recognises the Pleistocene as ending 11,764 years ago, because that is when the last advance of the ice—known as the Younger Dryas—ended. After that comes the shortest geological epoch of all: the Holocene.

I hate to quibble, but if glacial cycles characterise the Pleistocene, then we are (or were until a few decades ago) still in it—for the simple reason that the ice would have advanced again, in accordance with the Milankovich cycles. But over the past 20 years or so, the burden of greenhouse gases has built to such an extent, and the planet so warmed, that scientists are confident that the ice will not return.

A proposal currently before the International Union of Geological Sciences argues for the recognition of yet another geological period: the Anthropocene. It is defined as commencing at the moment that human activity began to leave an indelible and widespread stamp on Earth's sediments. Perhaps the moment that our greenhouse gases prevented the future return of the ice is the appropriate marker. Under that reading, the Pleistocene should last until around the end of the twentieth century, when it was succeeded by the Anthropocene.

The Pleistocene is characterised by rapid shifts in climate, including eleven major glacial events—ice ages—along with many minor ones. On each occasion glaciers and ice sheets spread and remained for extended periods, before being melted by brief warm spells. Over the Pleistocene, ice ages have prevailed for 90 per cent of the time, and at their greatest extent, glaciers and ice covered 30 per cent of the surface of the Earth. In the northern hemisphere, permafrost or glacial desert extended for hundreds of kilometres south of the ice sheets. With a bit more cooling, it's possible that the glaciers could have extended to the equator.*

These dramatic climatic shifts left behind much evidence, in the form of glacial features and altered rainfall patterns. But it was not

---

*Just why the ice did not grow and grow until it covered the Earth is still debated. One factor may have been the polar deserts, the dust from which may have coated the ice, dulling it. The oceans played an important role in accelerating the weak warming trend triggered by the celestial cycles because warm water holds less gas than cold water.

until 1837 that the Swiss scientist Louis Agassiz (he who named the megalodon shark) introduced the idea that much of the Earth had recently been in the grip of ice. He migrated to the US in 1847, taking up a position at Harvard, and in New England discovered abundant evidence, including massive boulders shifted by the ice, to support his hypothesis. But just what caused the ice ages remained a mystery until Milutin Milanković, a Serbian mathematician and a highly successful civil engineer, turned his mind to the problem.

Born on the banks of the Danube in what is now Croatia, Milanković began researching the causes of the ice ages in 1912. But, what with building bridges and experimenting with concrete, he lacked the time to make much progress on matters celestial. When World War I broke out, Milanković was on his honeymoon in his natal village, where he fell foul of the complex and shifting politics of Eastern Europe of the time. Considered a hostile foreign national, he was arrested by the Austro-Hungarians and taken to Esseg fortress, where he was locked up as a prisoner of war. He wrote that:

> The heavy iron door closed behind me…I sat on my bed, looked around the room and started to take in my new social circum-stances…In my hand luggage which I brought with me were my already printed or only started works on my cosmic problem; there was even some blank paper. I looked over my works, took my faithful ink pen and started to write and calculate…When after midnight I looked around in the room, I needed some time to realise where I was. The small room seemed to me like an accommodation for one night during my voyage in the Universe.

Mrs Milanković, it appears, was less sanguine about the arrange-ment. Through a colleague in Vienna she organised for Milutin to accompany her to Budapest. And there, through the good graces of other colleagues, he was given the run of the libraries of the Hungarian

Academy of Sciences and the Hungarian Meteorological Institute. Milanković spent almost the entire the war happily studying the climates of other planets, as well as the great problem of the ice ages, and in the fragile peace that followed he became a professor of mathematics in Belgrade.

In 1930 Milanković published research showing that the ice age was caused by slight variations in the Earth's orbit around the sun and the tilt and wobble of the Earth on its axis. By mid-1941 he had finished a book explaining his full theory, *Canon of Insolation of the Earth and Its Application to the Problem of the Ice Ages,* which included an explanation of the trigger for the advance of the ice: when celestial factors led to cool summers in the northern hemisphere, not all the winter snow would melt; year by year the ice caps would grow, and because ice is very bright and reflects sunlight, it accelerates the cooling trend.

On 2 April 1941 Milanković delivered his manuscript to a printing house in Belgrade. Just four days later, disaster struck when Germany attacked the Kingdom of Yugoslavia and destroyed the printery in a bombing raid. What one war had given, a second threatened to take away. But, thankfully, some printed sheets survived in a warehouse. A month later, in May 1941, two German officers called on Milanković, bearing greetings from Professor Wolfgang Soergel. They explained that they were students of geology, and Milanković entrusted them with the only remaining complete copy of his work. Soergel ensured that the book was published—in German—but in the decades after the war Milanković's *Canon* was ignored. When the first English translation was made in 1969, it revolutionised our understanding of the ice ages immediately.

The cycles that Milanković identified had been in existence for hundreds of millions of years. So why did they trigger an ice age beginning 2.6 million years ago? It seems that, in earlier times, the configurations of the continents, and the higher level of greenhouse gases present in the atmosphere, prevented a full chilling, regardless of the influence of Earth's orientation relative to the sun. From about 2.6 million years ago, however, these buffering effects were removed, and Milanković's cycles began to play Europe's biota like a piano accordion. At first, the duration of each cycle was about 41,000 years, and the effects were gentle. But about a million years ago the cold spells (known as glacial maxima) got deeper and longer, the cycles extending from 41,000 to 100,000 years.[1] Just why this shift—from 41,000- to 100,000-year-long cycles—occurred is hotly debated. But the impact was clear: two faunas started to develop across Eurasia; a new one adapted to the cold phase, and an older one acclimated to the warmth.

The ice ages were not kind to the warmth-loving fauna. As the cycles intensified they blew entire species from the cosy café of temperate Europe. At each contraction of the bellows, frigid northern winds blew outwards from the pole, forcing the warmth-loving elements of its flora and fauna into shrinking refuges in Spain, southern Italy and Greece, where they would be confined until the orbital patterns ushered in a brief warming. The European ice age is thus marked by migration and extinction on a massive scale. More than half of Europe's mammal species disappeared with the onset of the ice ages; surviving was all about adaptation, and migration.

What was it like to live in ice-age Europe? During the last glacial maximum, which peaked about 20,000 years ago, the sea level was between 120 and 150 metres lower than it is today, courtesy of all the water locked up as ice. A broad plain was exposed across the north of Europe, connecting Ireland to Britain and the continent. To the north a great field of ice and snow extended across land and sea to the pole. In the south, only the shallower northern Adriatic was exposed, though some of the Mediterranean islands became connected (Sardinia with Corsica, and Sicily with the mainland, for example). Sea temperatures were as much as 13° Celsius lower than they are today, and the now-extinct great auk bred on the coast of Sicily, while gulls, auks and gannets nested by the million on the Mediterranean cliffs of Iberia, France and Italy.

On land, temperatures were probably 6–8° Celsius lower, on average, than those of today, and winters were much colder, with permafrost extending as far south as Provence. Strong winds blew from the high ice cap, carrying fine dust from the polar deserts throughout Europe. Where London, Paris and Berlin are today, a vast polar desert, all but devoid of plant life, extended to the line of ice on or beyond the northern horizon. Frostbite, grit in the teeth and lungs full of dust would have been the lot of anyone venturing so far north.

To the south of this frigid desert, in a band stretching from northern Spain to northern Greece, were steppe lands and a stark forest of taiga-like conifers grew, similar to that which covers areas of Siberia today. Further south, deciduous trees and Mediterranean scrublands (maquis) found a refuge. Though limited in extent, these warm-adapted habitats were remarkably diverse. Around Gibraltar, for example, one could walk in a pine or oak forest, harvest blueberries and stroll through the maquis now typical of the region, all in the one day.[2]

The cold periods that characterise the ice ages end abruptly, accelerated by the outgassing of CO_2 from the oceans when they begin to warm. But it takes thousands of years for the climate to reach a new, warmer equilibrium. At the end of the last glacial maximum it took the ice between 12,000 and 13,000 years to melt, and for the sea to regain its current levels. Before that, the Black Sea was a freshwater lake, with people living along the coasts, which were about 150 metres below current sea levels. Then, 8000 years ago, the Mediterranean broke through the Dardanelles and the Bosphorus, and within a few years the Black Sea filled, displacing those living around its ancient shore. The melting of the ice released weight on lands across Europe.

Some areas, including Basilicata in southern Italy, the Gulf of Corinth, and northwestern Scotland, were uplifted by several hundred metres. There are many strange consequences of this history. One is evident if you are a birdwatcher standing in a mature Mediterranean forest. You will not hear a single species unique to the Mediterranean region. Yet in the nearby maquis you can hear plenty. That's because, even as far south as the Mediterranean, tall forests were so devastated by the ice ages that none of the surviving patches was large enough to support the species of birds that were restricted to them.[3]

Between 2.6 million and 900,000 years ago, when the ice-age cycles were 41,000 years long and relatively gentle, a characteristic fauna developed. The giant hyena, *Pachycrocuta brevirostris*, was a metre high at the shoulder and 190 kilograms in weight, making it the largest hyena that ever lived. It evolved in Africa and its earliest appearance in Europe dates to about 1.9 million years ago.[4] The giant hyena used caves as den sites, and the remains of past meals are well preserved in some. Even the bones of large creatures like hippos and rhinos bear its distinctive chew marks, though whether the hyenas killed such beasts or merely scavenged their carcasses is not known. Giant hyenas were probably social and were certainly powerful enough to kill creatures the size of wisent, or perhaps to drive *Homo erectus* from its caves.

The giant hyena arrived from Africa at about the same time that our ancestors, *Homo erectus*, reached Europe. The hyenas thrived, but our ancestors remained rare almost to invisibility. But then, about 400,000 years ago, the great hyenas vanished, and new members of our genus, in the form of early Neanderthals, started to become abundant, and to use caves.[5,6] What caused the giant hyena to become extinct is not clear. But some researchers tie it to the decline of the sabre-toothed and scimitar-toothed cats, as giant hyenas scavenged the big cats' kills. The European sabre-toothed cat (a relative of the *Smilodon*) had become extinct about 900,000 years ago, while the giant scimitar-toothed cat, *Homotherium*, had begun to decline in Europe by half a million years ago.

The European jaguar inhabited the continent from about 1.6 million years ago until around 500,000 years ago. Larger than the living jaguar of South America, it is sometimes considered to be a giant version of the South American species that was replaced in the Old World by the leopard. Another spectacular cat of the early European ice age was the giant cheetah: it was the height of a lion, though considerably lighter. It had disappeared from Europe by about a million years ago.

Early ice-age Europe was also home to giant beavers of the genus *Trogontherium*. At almost two metres long they had similar gnawing habits to today's beavers, but lacked flattened tails, instead having longish cylindrical ones. They survived in parts of Russia until about 125,000 years ago. Europe's giant beavers lived at the same time as the first moose, *Libralces gallicus*. Its two-million-year-old remains have been found in southern France, where it inhabited warm grasslands.

An unexpected African immigrant was a kind of hippo, *Hippopotamus antiquus*. It had arrived by 1.8 million years ago and was happily settled into the Thames, among other rivers, by the time a warm spell known as the Eemian occurred between 130,000 and 115,000 years ago.* At the time, temperatures briefly rose to become slightly higher than those of the pre-industrial period, making the Eemian the warmest time in the last million years.

A small, ancestral red deer had appeared in Europe by about two million years ago.[7] Its leg bones suggest that it may have been adapted to rugged mountain environments. It shared the forests with an early form of fallow deer. By a million years ago larger red and fallow deer had evolved, that were very similar to living types. Another group that flourished during the early ice ages was the ancestral cattle, bison

---

* The Eemian is also known as Marine Isotope Stage Five E.

and muskox, and an ancestral form of the giant deer *Megaloceros*.[8] By 900,000 years ago—just as the 100,000-year glacial cycle takes hold—the ancestors of the cave lion, the first lions to be seen in Europe, stalked into the continent.

Ancestral wolves, *Canis etruscus*, had arrived in Europe from Asia over three million years ago, but they did not flourish until the ice ages. They can survive in many habitats, but wolves are really at home on the tundra.[9] In European fossil deposits their bones are often accompanied by the remains of a coyote-sized canid, *Canis arnensis*. Over time this smaller canid became restricted to the lands adjacent to the Mediterranean, until about 300,000 years ago when it died out in Europe. Today, dogs may be our best friends, but strangely, in the whole fossil record of Europe there is only one site where a primitive human-like creature and a primitive wolf co-occur: the 1.85 million-year-old Dmanisi site in Georgia.

CHAPTER 23

Hybrids—Europe, the Mother of Metissage

Advances in DNA analysis, particularly in the study of ancient DNA, are unlocking a hitherto unsuspected aspect of hybridisation (or *metissage* as the French might call it). It is increasingly shown to have been important in the origination of species, and in helping species adapt, with many examples coming from Europe. But perhaps most strikingly hybridisation has been a very important influence on human evolution in Europe. We often think of hybrids as something inferior—a sort of bastard or mongrel type. Pejorative associations of the word 'hybrid' were particularly common in the first half of the twentieth century, when misguided ideas about genetics made purity of race a dangerously attractive concept. The pioneering geneticist R. A. Fisher—who was a keen promoter of eugenics (the idea that societies could be improved by selectively breeding 'superior' humans)—believed that hybrids resulted from 'the grossest blunder in sexual preference which we can conceive of any animal making'.[1]

The idea that species are discrete entities—carriers of a unique genetic heritage, is deeply embedded within us, perhaps reflecting some sense of a perfect, pre-human world, so hybrids can threaten our sense of order. They certainly complicate the work of the taxonomists, some defying easy classification and threatening the Linnaean system of classification that has ruled biology for more than 250 years.

Yet we have long known that hybridisation is widespread. By 1972, about 600 kinds of mammal hybrid had been identified (many from zoos or other captive situations).[2] By 2005 it was estimated that 25 per cent of plant species, and 10 per cent of animal species, were involved in hybridisation.[3] Over the last few years, research into ancient DNA has revealed that such figures are gross underestimates, even for wild species living in nature. Two recent studies, one involving bear species and the other elephants, illustrate what is being learned.

The six bear species living today (polar, brown, Asiatic black, American black, sloth and sun) have evolved from a common ancestor over the last five million years. While they remain very different in appearance and ecology, DNA analysis reveals an astonishing degree of hybridisation in their lineage. For example, polar bears have hybridised with brown bears, so that 8.8 per cent of the brown bear genome comes from polar bears. This means that 'Pizzlies' (as recent polar bear/brown bear hybrids are known), are not a new phenomenon, but have been produced for hundreds of thousands of years. Among many other crossings noted in the study, brown bears have hybridised with American black bears, and Asiatic black bears with sloth bears, and sloth bears with sun bears. The researchers concluded that hybridisation between various bear species has been going on for millions of years, so that when the crosses between bear species are included on the bear family tree, the diagram looks more like a family network.[4]

The history of hybridisation among elephants is, if anything, even more astonishing. A recent study by the Harvard-based palaeogeneticist Eleftheria Palkopoulou and her colleagues, which includes the three living species (African, African forest and Asiatic) and three extinct kinds (European straight-tusked, woolly mammoth and American mastodon), revealed that elephants have hybridised throughout most of their history. Indeed, some extinct elephants result from such extensive hybridisation that they are not easily classifiable in the Linnaean system.

In summarising their findings, Palkopoulou's team say: 'The capacity for hybridisation is the norm rather than the exception in many mammalian species over a time scale of millions of years.'[5] They also speculate that the sharing of genes through hybridisation might have helped species migrate and adapt to threats and opportunities by allowing species to acquire genes from near relatives. Looked at this way, we can think of species, such as our own, that have now lost the ability to hybridise because our close relatives are extinct, as vulnerable and isolated.

Were hybridisation extensive enough, life would become one undifferentiated mass. So why do species exist? It turns out that there are mechanisms (known as species isolating mechanisms) that make the production of hybrids difficult. It is rare for an individual to overcome these barriers, but among the millions of individuals that comprise a species it is common for enough hybrids to be produced to allow genes to flow between species. Some species' isolating mechanisms are behavioural—such as the possession of a particular mating call, which only females of a given species will respond to—or a preference for reproducing at a particular time of the year. Others, such as penis size or shape, are physical. But there are also genetic and epigenetic barriers. Sometimes genetic factors prevent a viable embryo from forming. But they can also result in most first-generation hybrid individuals being infertile, or having low fertility. In a phenomenon known as Haldane's rule, this is particularly true for male hybrids among mammals. But if first-generation hybrids do manage to produce some offspring, the next generation often has improved fertility—though usually only with one or the other species (but not both) involved in the original cross. All of these barriers tend to limit, but not eliminate entirely, the flow of genes from one species to another.

Sometimes hybridisation does more than permit gene flow between species, instead creating an entirely new, hybrid species. Among the

European species that arose by hybridisation is the European edible frog (*Pelophylax kl esculentus*); a widespread and economically important creature esteemed as a culinary delicacy in France. The parent species that gave rise to it—probably hundreds of thousands of years ago—are the European marsh frog and the European pool frog. As you might have noticed, the creature's scientific name has a 'kl' inserted in it. This denotes that it is a 'klepton' or 'gene thief'—a hybrid that requires another species to complete its reproductive cycle for it. Most kleptons are female, and some don't use the genes of the male at all, merely deploying his sperm to stimulate the egg into development without fertilising it.[6]*

Even some mammal species arose via hybridisation. The golden jackal has recently been recognised as being two different species—a smaller one that originated with an early offshoot of the wolf lineage, and a larger kind that is closer to the modern Eurasian wolf and whose ancestors must have migrated into Africa before mixing with the smaller jackal and creating a new, hybrid species.**

The wisent—Europe's largest surviving mammal—is a stable hybrid species that arose around 150,000 years ago, when aurochs and steppe bison underwent an extended period of hybridisation. Steppe bison (from which the American buffalo is descended) inhabited the mammoth steppe and vanished from Europe at the end of the last glacial period, while aurochs were creatures of the more temperate forested areas. Wisents carries mostly bison genes, with a healthy infusion (about 10 per cent) of aurochs genes, the mixed genetic heritage

---

* The genetics of kleptons can be extremely complex, some eliminating the genes of one parent during the production of sperm or eggs. Three klepton hybrid species, all with the marsh frog as a parent, exist in Europe, and all have distinct genetic pathways for reproduction. In all three, the genes of the marsh frog are never lost.

** Neither of these species is closely related to the 'true jackals' of the genus *Lupulella*.

apparently helping the wisent survive changing conditions as the climate warmed and forests spread.[7]

Hybridisation in agriculture is different from hybridisation in nature, both because the conditions created by people allow hybridisation between species that would never hybridise naturally, and because we have selected for extreme traits in many domesticated forms. When domesticates become feral, or breed with non-domestic relatives, those interested in conservation face a quandary: should they seek to eliminate the hybrids, and so seek to ensure that the wild types are not overwhelmed by the domesticates? Some see the highly modified domestic creatures as a form of pollution—albeit genetic pollution—that because of the huge abundance of domesticates, can endanger their far rarer wild relatives.

For example, a case might be made that dog-wolf hybrids should be removed from nature for fear that dog genes might overrun the wolf population (an issue I shall return to). But a more difficult example involves the Scottish wild cat, in which the great majority of the population are hybrids between wild and domestic types. It might seem desirable to remove the hybrids, but to do so would leave a population so small as to be headed for extinction.

Hybrids present a particular problem when it comes to legal policy. Our major legal instruments for protecting species, including the Bern Convention and the US Endangered Species Act of 1973, deal with species, not hybrids. Indeed, the US Act has been described as 'almost eugenic' because it excludes hybrids from protection.[8] Given our knowledge of the extent of hybridisation, this is problematic. And matters are made more difficult because defining hybrids is not always straightforward. A first-generation cross might stand out, but over time it becomes more and more difficult to detect hybrid animals. Indeed, most of our recent insights into the importance of hybridisation in nature come from DNA studies on animals that, at first glance, appear not to be hybrids.

Hybridisation can also result in heterosis—the scientific term for the production of 'super fit' hybrid individuals—many examples of which come from agriculture. Heterosis can be thought of as the opposite of inbreeding depression, the phenomenon whereby the offspring of individuals that are genetically too similar—for example, brother and sister—can suffer from debilitating maladies. Heterosis usually occurs when the parents are moderately different, for if individuals are too different, their genes often will not combine to form a viable embryo. Heterosis is well known to animal and plant breeders, who seek it out: grains that result from crossing different strains, for example, are often more disease resistant and grow faster.

An instructive example of a heterotic individual is 'The Toast of Botswana'. The Toast is a cross between a female goat and a male sheep. As such, he is an exceedingly rare beast—goats and sheep being too different genetically to readily create a viable offspring. The Toast was born into the flock of Mr Kedikilwe Kedikilwe at the Botswanan Ministry of Agriculture, who noticed that the creature grew faster than the lambs and kids born at the same time. He was also astonished by the fact that it hardly ever got sick, even when an outbreak of foot and mouth disease afflicted the rest of the flock.

As its name suggests, for a time after its birth The Toast was exemplary in every way. But when he reached puberty a problem arose: the creature became extremely libidinous, copulating with sheep and goats indiscriminately and even out of breeding season. This unseemly behaviour earned him the shameful nickname of Bemya, or 'rapist'. Yet, despite his ceaseless efforts, The Toast fathered no young. Ashamed and annoyed at its fall from grace, Mr Kedikilwe had The Toast castrated.[9]

Hybrids are often noted for their libidinous ways—as if they understand that the only possibility for them to pass on their genes lies in indulging in many and varied couplings, in the hope that some way around the species isolating mechanisms can be found. But because we humans misapply moral standards to animals, we often cut short their efforts. Had Mr Kedikilwe stayed the knife, we may have learned a great deal more about heterosis and hybridisation.

Heterosis can affect far more than growth rates and disease resistance, for brain function and behaviour can also be influenced, as is evidenced by the mule. As Charles Darwin observed, the mule 'always appears to me a most surprising animal. That a hybrid should possess more reason, memory, obstinacy, social affection, and powers of muscular endurance than either of its parents, seems to indicate that art has here outmastered nature'.[10] We consider some of the mule's key traits as observed by Darwin—reason, memory and social affection— as among our species' most valued and distinctive characteristics: yet we never consider that they may result from heterosis.

Because of its position at the crossroads of the world, Europe has had many immigrant species that provided unprecedented opportunities for hybridisation. It may be this fact, as much as anything, that has driven evolution at such a rapid pace in Europe, and which in turn has lent many European species the capacity to colonise new and environmentally different lands. The pace of hybridisation in Europe has picked up substantially since the dawn of agriculture, and ever more hybrid species are being created. The Italian sparrow, for instance, is a hybrid between the Spanish sparrow and the house sparrow that originated in Italy sometime in the last 10,000 years.[11] In Britain alone, at least six new species of plants have arisen through hybridisation since 1700, while hybrid super-slugs are becoming a plague in English gardens.[12,] As climate change brings evermore creatures to Europe, the rate of hybridisation is likely to skyrocket.

The idea that hybridisation may be 'the norm' among mammalian species for millions of years after they first arise—and that it may help them adapt to new conditions—is very challenging to many, and is diametrically opposed to the idea that they result from nature's 'grossest blunder'. But Fisher's views of hybridisation are now as outmoded as his endorsement of eugenics. It's now clear that species are not 'fixed' entities, but are permeable. Throughout European prehistory, immigration has created opportunities for heterosis to arise in the wild, and as a result European nature is all the better adapted. Perhaps in time we will come to value many hybrids, and to see that there can be no more dangerous concept than the idea of racial or genetic purity. At a minimum, our new understanding of hybrids means that a fundamental re-think of classification, endangered species legislation and laboratory-based gene transfer is overdue.

CHAPTER 24

Return of the Upright Apes

Between 5.7 million years ago, when a small ape strolled the seashore in what is today Cyprus, and 1.85 million years ago (when *Homo erectus* appears), there is no evidence of apes in Europe. Our lineage had been evolving in Africa, and the creatures that returned to Europe belonged to our own genus—*Homo*. Everything we know about them comes from a fossil site at Dmanisi, in Georgia, where a rich collection of *Homo erectus*, along with many other species, was found in the 1980s.[1]

Located on a promontory-like plateau overlooking the confluence of the Pinasauri and Masavera Rivers some 85 kilometres southwest of Georgia's capital Tbilisi, the deposits are preserved under the medieval ruins of Dmanisi, which was taken from the Turks and rebuilt by the Georgian King David the Builder in the twelfth century. The bones are preserved in gullies that were cut into the plateau and subsequently became filled with sediments. In 1984 a team began a major excavation resulting in the discovery of an abundance of stone tools and hominid remains. Work at Dmanisi continues under the direction of David Lordkipanidze, Director of the Georgian National Museum, with new discoveries occurring every few years.

Dmanisi has forced a rethink of both human and European prehistory. The deposits are between 1.85 and 1.78 million years old, making the *Homo erectus* remains found there the oldest known.[2] The brains of the Dmanisi *Homo erectus* are 600 to 775 cubic centimetres in volume (about half the size of the brains of anatomically modern humans). This is much smaller than those of other *Homo erectus*, and closer in size to *Homo habilis* (the African ancestor of *H. erectus*). One extreme view is that *Homo erectus* evolved in Europe from some earlier, as yet undetected species of *Homo*. Whatever the case, it is striking that, below the neck, the Dmanisi *Homo erectus* are similar to modern humans, though their arms retained some primitive characteristics typical of more arboreal ancestors.[3] Another striking feature of the Dmanisi remains is their variability. Both large and very small individuals are included; palaeoanthropologists posit that if the five skulls recovered to date had been found in separate locations, they would be classified as belonging to several different species.

A toothless cranium of a male *Homo erectus* found in 2002 matched up perfectly with a toothless jaw found in 2003, and the discoveries open a window on the social life of the species. In many other animals, a lack of teeth means death: the individual starves. The toothless *Homo erectus* from Dmanisi provides the oldest evidence anywhere for the survival of such an impaired individual. Lordkipanidze argues that the man could only have survived with help; the Dmanisi *Homo erectus* must have been highly social, perhaps living in small family groups that cared for its less able members.[4]

Could *Homo erectus* speak? Rarely preserved parts of the cranium and spine unearthed at Dmanisi (including a string of six vertebrae) are shedding some light. The Dmanisi *Homo erectus* had the right respiratory apparatus to support speech—indeed it is within the range of our own species.[5] And an enlarged depression on the inside of the cranium provides evidence that Broca's area, the part of the brain that processes articulated language, was present, so it's possible that language was used by the bipedal apes of Dmanisi.

Many palaeoanthropologists would reject the idea that *Homo erectus* had language, viewing the palaeontological data as akin to building a castle on sand. But we should be cautious; ever since Victorian gentlemen imagined the Neanderthals as lowly cave men and themselves as the acme of evolution, we have underestimated the capacities of our distant ancestors and relatives. And, with each new scientific finding, we're discovering them to be more competent than we had previously thought.

The *Homo erectus* of Dmanisi were capable predators who repeatedly occupied the plateau for at least 80,000 years. It must have been a valuable strategic vantage point from which to watch for migrating animals. Fossilised hyena faeces and the bones of fourteen other carnivore species reveal that *Homo erectus* did not have sole tenure over the lookout. We can imagine a European Serengeti moving through the river valleys, with the predators descending from their lookout to kill, then transport the flesh to the hilltop to be eaten. The transported remains of elephants, rhinos, giant ostrich, extinct giraffes, seven species of antelope, goats, sheep, cattle, deer and horses have been found, the last two being especially abundant.[6]

Just how the various predators interacted can only be guessed at. It seems likely, however, that the giant hyena and *Homo erectus*, being the largest and most social species, tussled for control of the lookout. While the hyena was substantially larger than *Homo erectus*, the hominids had the advantage of tools such as projectiles. I suspect that more often than not, in an open site like Dmanisi, *Homo erectus* (a diurnal ape of tropical origin) would have come out on top of the nocturnal hyenas. In the dark of caves, however, the tables would almost certainly have been turned.

For about a million years after the Dmanisi individuals lived, evidence of *Homo erectus* is exceedingly rare in Europe. But we know from fossils preserved elsewhere that the species was becoming bigger-brained and was developing a more diverse toolkit. The next clear glimpse we get of Europe's upright apes comes from the Sierra de Atapuerca in northern Spain, where caves have yielded fragmentary bones and tools dating to between 1.2 million and 800,000 years ago. The most important site, at Gran Dolina, provides strong evidence of cannibalism. The remains are mostly of juveniles, and they bear butchering and tooth marks.[7] In 1997 these remains were named *Homo antecessor*. A few adult teeth and stone tools, dating to about 700,000 years ago, that were discovered in 2005 in cliffs at Pakefield, Suffolk, have been attributed to this species. Whether *Homo antecessor* is in fact just another form of *Homo erectus* remains an open question. I will be conservative and refer to it and all similar aged European remains as *Homo erectus*.

The discovery of *Homo erectus* remains in caves in Spain and Britain raises the issue of control of fire. Caves are cold and dark places that large carnivores prefer as lairs. The association of carnivores, but not most herbivores, with caves presumably relates to the amount of time an individual can stay under shelter. Carnivores kill infrequently, and spend days sleeping off their meal, while herbivores must spend most of their time foraging. Thus, except for hibernating species like cave bears, herbivores cannot benefit to the same extent from the more clement conditions provided by caves.

In ice-age Europe, the ability to commandeer caves was probably key to the survival of upright apes. Of tropical origin, they lacked insulating fur and could not survive freezing conditions without shelter. But competition for caves must have been fierce, and control of fire may have been decisive in allowing upright apes to retain a toe-hold as Europe chilled. The earliest evidence of the human use of fire is at best ambivalent, coming from burned sediment dating back 1.5 million years. There is better evidence that *Homo erectus* was using fire 800,000 years ago, and by half a million years ago (as evidenced by charred bones) some upright apes were cooking their food. But we should not assume that the discovery of hominid bones in caves confirms occupation.

It's possible that giant hyenas or cave lions carried the remains of *Homo erectus* into their lairs, or that they were washed in by floods.

The discovery in 2013 of footprints at Happisburgh, England, reminds us of how little we know about our human lineage in early ice-age Europe. The prints were made by a group of five individuals varying in height between 0.9 and 1.7 metres—possibly a family walking upstream along the estuary of the Thames, between a million and 780,000 years ago.[8] They may have left an island where they had spent the night in relative safety and were searching for food. Shortly after these astonishing footprints were documented they were destroyed by a high tide.

At the time these *Homo erectus*-like creatures wandered along the ancestral Thames, the climate of that part of Europe was cool—similar to that currently experienced in southern Scandinavia. The remains of a primitive mammoth and bison bones that bear signs of being butchered by humans have been discovered at Happisburgh. Perhaps the *Homo erectus* that left the footprints migrated north seasonally to hunt. Whatever the case, it is difficult to imagine our ancestors surviving year-round in such a climate without fire. As conditions chilled further *Homo erectus* vanished altogether from Britain, and probably from all of northern Europe, though they may have found refuge in the temperate peninsulas of Iberia, Italy and Greece.

When the Happisburgh footprints were made, the 100,000-year ice-age cycle had commenced. Each advance of the ice was a little different from the ones that preceded it. The most extreme glaciation occurred between 478,000 and 424,000 years ago. Known as the Anglian glaciation in the UK, the Elsterian on the northern European mainland, and the Mindel in the European alps, it saw the ice reach as far south in Britain as the Scilly Isles.* In eastern Europe this glacial advance

---

* A universally recognised scientific name for the event is 'Marine Isotope Stage 12'.

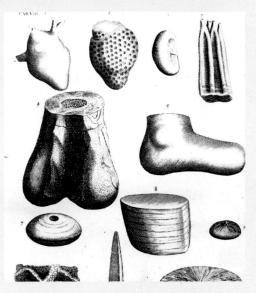

Robert Plot's 1677 depiction of the first dinosaur bone to be described (middle row, left). Later, the fossil would be named the *Scrotum humanum*.

A pig-nosed turtle. Once widespread in Europe, today the only surviving species inhabits New Guinea and northern Australia.

Franz Nopcsa von Felső-Szilvás, Baron of Săcel and discoverer of Europe's dwarfish dinosaurs, dressed as an Albanian Shqiptar warrior, 1913.

A 47-million-year-old fossil shark, nearly two
metres long, from Monte Bolca, preserving dark
pigment on the fin tips.

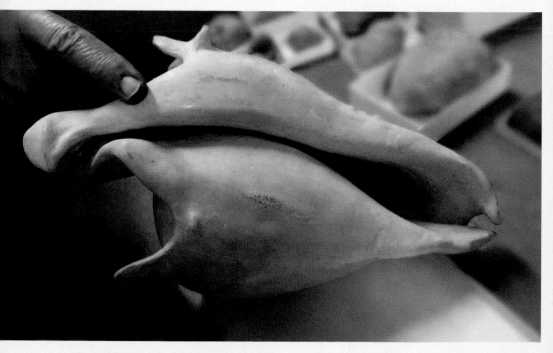

Gisortia gigantea, the largest cowrie ever, lived
about 50 million years ago in the Tethys Sea.

A fossilised giant bell-clapper shell—one of the largest gastropods ever to live. It thrived in the Paris Basin about 50 million years ago, and today its only surviving relative is found in the waters around southwestern Australia.

A model of a fossil *Nummulite*. Randolph Kirkpatrick of the British Museum believed that the entire planet was made of fossil *Nummulites*.

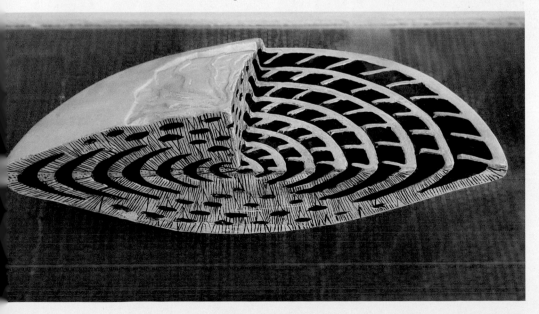

A nipa palm in the Solomon Islands. Forty million years ago, nipa palms grew around the European coast.

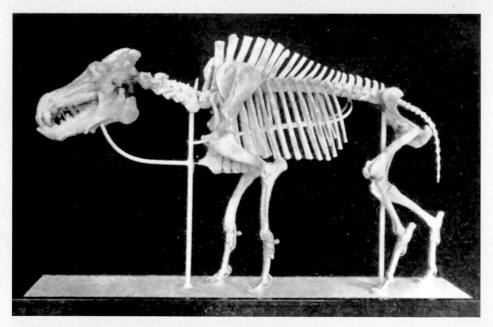

The skeleton of an entelodont. These 'pigs from hell' were top predators in Europe 30 million years ago.

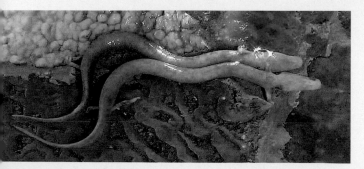

The olm, described by Johann Weikhard von Valvasor as 'a worm and vermin', is a cave-dwelling amphibian from the Slovenia region.

The skull of a deinothere. These elephants reached Europe from Africa about 16.5 million years ago and thrived there until 2.7 million years ago. Precisely how they used their strange tusks remains a mystery.

The skull of a *Hoplitomeryx*. These strange deer had up to five horns, and inhabited the now-vanished Mediterranean island of Gargano.

A skeleton of the Tuscanian ape *Oreopithecus bambolii*. An inhabitant of Sardinia eight million years ago, it appears to have been bipedal.

Sir Richard Owen, president of the British Association for the Advancement of Science, described the giant viper *Laophis* in 1857. He was one of the most dastardly scientists ever to live.

A cave bear skeleton. A gigantic vegetarian, with a skull up to three-quarters of a metre long, it was unique to Europe until its extinction 28,000 years ago.

A model of a
Neanderthal woman.
Constructed in 2014,
it can be seen in
the Museum of the
Confluences, Lyon.

A giant deer. With an
antler spread of more
than three metres, it
was a striking member
of the European
megafauna. It survived
on the Isle of Man until
about 9000 years ago.

The lion-person of Hohle Fels, Austria. Carved from ivory about 40,000 years ago, this imaginary hybrid creation is the work of the earliest human–Neanderthal hybrids.

A skull of the now-extinct scimitar-toothed cat *Homotherium*. This creature could weigh up to 440 kilograms. It survived in Europe until 28,000 years ago.

Konik ponies in the rewilded Oostvaardersplassen, Netherlands. With an area of 60 square kilometres, the Oostvaardersplassen, though managed by people, offers a vision of ice-age Europe.

appears to have caused the extinction of those venerable members of the frog and toad group, the palaeobatrachids, which we first encountered on Hateg. Their favoured habitats were large permanent lakes. With the extreme glaciation their last redoubts, in the valley of the Don River, in what is now Russia, became too dry for them.[9]* During the Anglian glaciation, the ice caps were smaller than during earlier glacial events, but conditions in the periglacial areas around the ice were far more severe. The last of the palaeobatrachids were squeezed out of existence between a severely cold periglacial north, and a desertified south. I must admit that missing out on seeing these marvellous and antique creatures by the merest sliver of time is an immense frustration!

The Anglian glaciation doubtless drove *Homo erectus*, its competitors such as the giant hyena, and its prey, from much of Europe. After the ice finally receded, new types of creatures would move north from Africa; the spotted hyena would replace the giant hyena, and a new kind of upright ape would enter Europe. Genetic analysis indicates that the Neanderthals evolved in Africa 800,000 to 400,000 years ago. They may have displaced the *Homo erectus* lineage; or they may have hybridised with them.** Whatever happened, we see no more of *Homo erectus* in Europe after about 400,000 years ago.

---

* In eastern Europe, the Anglian glaciation is known as the Oka glaciation.

** A minority of scientists argue that *Homo neanderthalensis* evolved from *Homo antecessor* in Europe.

CHAPTER 25

Neanderthals

The 'mammoth fauna', which is so evocative of ice-age Europe, first evolved during the Anglian glaciation, around the time the Neanderthals reached Europe; so that in our minds Neanderthals, mammoths and other ice-age fauna are forever associated. By 400,000 years ago some Neanderthals had pushed north into Europe and Asia, where eventually they spread as far east as the Altai Mountains, preying on mammoths, reindeer, horses and other species. The early Neanderthals (who existed 400,000 to 200,000 years ago) have been referred to variously as *Homo heidelbergensis*, *Homo erectus* or *Homo neanderthalensis*. I will call them early Neanderthals. They were slightly shorter than us, though their brains were about the same size as ours. Later Neanderthals, in contrast, had larger brains than those of people living today (though their bodies were larger as well). We tend to think of Neanderthals as primitive beings with a crude material culture. But six superbly crafted wooden spears discovered in a peat deposit near Schöningen in Germany and thought to have been made by early Neanderthals give the lie to this idea. Wooden tools do not generally fossilise well, so these spears provide a rare insight into Neanderthal wood technology. Made between 380,000 and 400,000 years ago, they were probably used to hunt horses. What is remarkable about them is their degree of sophistication. They are weighted towards the front and have finely crafted points: replicas performed as well as the best modern javelins, travelling up to 70 metres.[1]

Neanderthals also mastered the technology required to create adhesive pitch from tree bark. The earliest evidence was found in Italy and dates from between 200,000 and 300,000 years ago. This is long before *Homo sapiens* independently invented adhesives. Pitch manufacture requires foresight and the manipulation of materials and temperature (with the more sophisticated methods yielding far more than simple ones).[2] Researchers think that sophisticated methods were deployed in a manufacturing process requiring much preparation. Pitch is important in that it is used, among other things, to haft flint heads on wooden spears, creating highly effective weapons.[3]

A particularly rich haul of 5500 early Neanderthal bones, dating back 300,000 years and belonging to at least 32 individuals, has been recovered from the Sima de los Huesos site in the Atapuerca Mountains in Spain. The bones, many of which are from juveniles, were found at the bottom of a vertical shaft, where they form 75 per cent of all the remains found there, the rest mostly being ancestral cave bears and carnivores that may have been lured into the pitfall by the smell of rotting flesh. A single, beautiful red quartzite axe, made from materials sourced far away from the site was also found in the pit. Some researchers believe that the bones result from the disposal of corpses—a form of burial—and that the quartzite axe was a ritual offering to the dead.[4] If that's correct, it represents the oldest evidence found anywhere for care of the dead.

By about 200,000 years ago the 'classic' Neanderthal type—with its large nose, oversized brain and powerful body—had emerged. Neanderthals and *Homo sapiens* are extremely similar genetically, sharing 99.7 per cent of their DNA (by way of comparison, humans and chimpanzees share 98.8 per cent of their DNA). Because of this similarity, and the ability of humans and Neanderthals to interbreed, many writers refer to Neanderthals as humans. But doing so leaves us with no easy way to distinguish our own distinctive human type. So I will reserve the term 'human' for *Homo sapiens*.

The first Neanderthal remains to receive scientific attention were bones unearthed by quarry workers in 1856, in Feldhofer Cave in the Neander Valley, near Dusseldorf. They were passed on to savants, who made various suggestions about their identity. One thought that they were the last mortal remains of an Asiatic soldier who had died serving the czar in the Napoleonic Wars, while another thought that they were from an ancient Roman. Yet another identified them as belonging to a Dutchman.

In 1864, following publication of Darwin's *Origin*, the bones came to the attention of the geologist William King, then working at Queen's College, Galway. He described them, bestowing on them the name *Homo neanderthalensis*. Shortly after, King changed his mind, averring that the bones should not be placed in the genus *Homo* because they came from a creature that was incapable of 'moral and theistic conceptions'.[5] Despite his equivocation, King's name for the bones was published, and a good thing too, for the German biologist Ernst Haeckel was also studying the bones, and his suggested name for them was ghastly.

Haeckel was an extremely capable scientist who constructed the first comprehensive tree of life, named thousands of species, and coined such phrases as 'stem cell', and 'First World War'. But in 1866 he published the name *Homo stupidus* for the Neanderthal fossils, which—I cannot refrain from saying—reveals a certain lack of tact.[*] Under the rules of the International Code of Zoological Nomenclature, King's name *Homo neanderthalensis* (despite his second thoughts) has priority, and so it is the one used today.

---

[*] It is odd that Haeckel overlooked the very large Neanderthal brain, which was known from the original skull cap.

Most evidence of Neanderthal life comes from sites dating to the last 130,000 years, by which time the Neanderthals had become exquisitely adapted to the demanding European ice-age environment. With males weighing an average of 78 kilograms, and females 66 kilograms, analysis of the chemical composition of their bones reveals that they were obligate carnivores. Their refuse dumps show that their main prey were red deer, reindeer, wild boar and aurochs, though they occasionally tackled more challenging species such as young cave bears, rhinos and elephants.[6] In extreme circumstances they would, however, eat a little plant matter and fungus, as well as each other: twelve skeletons from El Sidrón Cave in Spain, bearing marks of death blows and defleshing, offering clear evidence of cannibalism.

Like many other carnivores, Neanderthals favoured caves as home sites, and were doubtless able to eject competitors from preferred lairs. There is ample evidence that they had mastered fire, and their tools indicate that they crudely prepared furs, perhaps to wear as cloaks, though they did not make fitted clothing. Their cave-dwelling habits, fire and cloaks were essential in allowing them to occupy much of Europe south of the ice.[7]

Genetic studies indicate that there were no more than 70,000 Neanderthals at any one time, and that they were spread thinly across all western Europe.[8] The genome of a female from Croatia has revealed low genetic diversity, as a result of existing as part of a small, isolated sub-population over multiple generations. One female whose remains were found in the Altai Mountains of Asia was highly inbred—a half-brother and sister being her parents—though this was not characteristic of all Neanderthal groups.[9] The bones of the dozen cannibalised individuals found in El Sidrón appear to be the remains of a family group that had been surprised, perhaps in their cave, before being killed and eaten. Forensic DNA analysis of their bones revealed that the males were closely related, but the females were not. This implies that

Neanderthals were similar to many recent and current human societies, in which the females leave their extended-family groups to marry
into other groups.[10]

Neanderthals were immensely strong, and many skeletons show
signs of injury consistent with mishaps incurred while hunting large
mammals with hand-wielded weapons. Despite their large brains,
their foreheads receded sharply, and their eyes were shaded beneath
pronounced bony brow ridges. They had barrel chests, which may
have helped them retain body heat, and large noses that were probably
useful for filtering ice-age dust, as well as warming the air they inhaled.
Just how hairy they were remains conjectural. Analysis of DNA indicates that their skin was pale, their eyes often blue, and their hair red.[11]

The eyes of Neanderthals were larger than ours, as by some measures were their brains.* In modern humans, we consider these positive
attributes. The question of Neanderthal brain size, however, has come
into dispute, one group of researchers arguing that a larger proportion
of the Neanderthal brain than our own was concerned with vision, and
that therefore less of it was involved with other functions. The same
study posits that Neanderthals were larger than modern humans, and
that therefore their brains were smaller relatively than ours.[12] Even if
this is so, we are left with irresistible questions: how did those large blue
eyes see the world, and what did that undoubtedly able brain make of
it? Alas, archaeology can only go so far towards answering them.

Did Neanderthals bury their dead? Sarah Schwartz of the
University of Southampton claims that their burial practices were
widespread. But the evidence she cites, including defleshing and the
concentration of bones in niches, could also result from cannibalism
or natural processes.[13] Whatever the case, a lack of complex burial

---

* Their large eyes may have been adaptations to the low light conditions of the
European winter, or to life in caves.

practices may not denote a lack of affection for the deceased. Among some African herders, a corpse was sometimes placed outside the thorn-bush fence surrounding the settlement. In the morning, the deceased had become new life, in the form of a hyena.

Neanderthal art at least 65,000 years old, and possibly much older, has recently been identified at three sites in Spain. Hand stencils, ladder-shaped designs and abstract shapes, all in red ochre, have been documented, but there are no depictions of animals.[14] Evidence of personal adornment is also scant, important exceptions being 118,000-year-old perforated and painted seashells from Spain, and 130,000-year-old white-tailed eagle talons discovered in a rock shelter in Croatia that had been modified in ways suggesting that they were strung on a necklace.[15] Somewhat more speculatively, a number of vulture wing bones found in caves on Gibraltar has led some researchers to believe that Neanderthals living there used vulture feathers as adornments.

The discovery in Bruniquel Cave in southwestern France of two ring-like structures (the largest 6.7 metres across) and six raised struc-tures made from about 400 large, carefully broken-off and stacked stalactites astonished scientists when it was published in 2016. All were built in a cavern which is in total darkness more than 300 metres from the cave entrance. The space must have been artificially lit, and there is abundant evidence for the use of fire around the stone circles.[16] Stalactites grow, so the moment they were broken off can be precisely dated—to 176,000 years ago, allowing no room for doubt that the work was done by Neanderthals. The purpose of the structures remains unknown; some speculate that they were the backdrop for some sort of ritual, while others think that they were merely part of a shelter. Whatever the case, they underline the fact that the Neanderthals were capable of great works and that much remains to be discovered about them.

Another aspect of Neanderthal culture is highly revealing of their inner lives. Neanderthals killed cave bears (often cubs) perhaps ambushing them as they emerged from hibernation. This could have been done from strategic points in cave systems, where it was possible to drive off the adults with fire or spears. Whatever the hunt methodology, Neanderthals have left extraordinary evidence of what has been dubbed 'the cult of the cave bear' throughout Europe.

One of the most striking examples was discovered in Romania's Altar Stone Cave in the Bihor mountains of Transylvania in 1984. Cavers from *Politehnică Cluj* explored the spectacular cave, whose vast chambers with their titanic stalactites and delicate cave ornaments pierce through an entire mountain. In his account of the discovery, Cristian Lascu writes of crawling, swimming and walking through the cave for a day and a night before reaching the site.

> The bear's cemetery made its sudden appearance before us, in a horizontal passage with vaulted ceilings with hanging tubular stalactites of impressive sizes. First, we saw a small skull, covered in popcorn concretions. Then two more, with long bones in front of the snout. Further away, in a depression of the floor, there was the skull of an adult bear measuring almost half a metre, and in a niche we found a mix of jaws, skulls and vertebrae. Next to this a large number of skulls belonging to young and adult bears were hardly visible under a thick layer of calcite. Four of them attracted our attention: they were arranged in a tight formation, with the occipital towards the interior, making a sort of imperfect cross.[17]

The arrangement of four juvenile cave bear skulls in a cross, along with limb bones placed in front of adult skulls cannot have been accidental. Similar finds have been made in other European caves and it is thought that the placement of a limb bone in front of the cranium, and the cross-like or back-to-back arrangements of juvenile skulls, which

are sometimes surrounded with pieces of flint, were part of an appeasement ceremony by Neanderthals.

Human hunters from many cultures have performed ceremonies
involving the skulls of bear species. After a successful polar bear hunt,
for example, the dead bear is treated with the greatest respect by Arctic
hunters. 'Don't be offended,' the Chukchi hunter says to the dead bear,
while the neighbouring Yupiit explain that they are only taking the
bear's muscle and fur, not killing it, for the soul of the beast lives on.
Elsewhere, gifts are presented to the skulls of slain bears—knives and
harpoon heads to males, and needles and beads to females.[18] In some
instances, 'altars' are set up, on which the bear skulls and gifts are laid
out. These arrays are similar to the juvenile bear-skull arrangements,
with their associated flint tools, left by Neanderthals.

There is a mystery surrounding these Neanderthal bear skull place-
ments. The near-perfect state of preservation of many of the skulls is
characteristic of individuals that have died during hibernation and
decomposed, undisturbed, in the cave. Skulls of hunted bears often
show cut marks or other damage absent on these skulls. So it seems
likely that the arranged bones are from bears that died naturally.
Appeasement may therefore have involved what the Neanderthals saw
as the cave bear family (including both living and dead individuals),
rather than just the individual they had hunted. If so, this reveals a
sophisticated comprehension of kinship.

The Neanderthals present a profound enigma. Although large-
brained and stronger than us, their material culture remained
rudimentary. It is striking that the great Neanderthal achievements—
including jewellery (dated to 118,000 and 130,000 years ago) and
stalactite structures (176,000 years ago)—are so very ancient. We've
discovered nothing like them from the last 80,000 years of Neanderthal
existence; yet the great majority of Neanderthal sites date to this later
interval. Did the Neanderthals suffer a sort of cultural simplification?

An informative parallel example can be seen in Tasmania's Aborigines. As explained by Jared Diamond in *Guns, Germs and Steel*, following their isolation from other Aboriginal groups as rising seas flooded Bass Strait about 10,000 years ago, the Tasmanian population of a few thousand lost the ability to make bone needles (and thus the ability to sew rugs) and possibly the knowledge required to make fire. If just one or a few individuals in a group know how to make or do certain things, the technology can be lost when they die. Genetic studies have confirmed that the Neanderthal population was small and fragmented. A loss of technologies over time may have resulted from isolation and small population size.

It must be said of both the Neanderthals and the Tasmanians that their capacity to innovate persisted. In the early nineteenth century Tasmanian Aborigines adopted dogs and guns following contact with the Europeans. And there is some evidence that once the Neanderthals made contact with humans they borrowed ideas and ways of doing things, in doing so creating the Châtelperronian culture, which persisted until the moment of Neanderthal extinction.

What to make of these most intriguing beings? We place such great emphasis on our own large brains in our claim to be *Homo sapiens*. Is it unreasonable to think that the Neanderthals may have exceeded us in some capacities? And what of their exquisitely made javelins, the equal of those that our best craftsmen can produce today, and of their ability to persist in the most extreme environments by hunting large, fierce prey? Imagine felling a woolly mammoth, or ousting a great hyena from its cave? I suspect that in some measures the Neanderthals were our superiors.

But zoogeography was against them. Africa is larger than Europe, and its tropical climate and the fertile soils of the Great Rift Valley make parts of it highly productive. This means that populations of large mammals were usually greater and denser in parts of Africa than in Europe. Moreover, modern humans seem to have occupied a broader ecological niche than the hyper-carnivorous Neanderthals, eating vegetable matter processed by cooking, which allowed humans to sustain higher population densities than Neanderthals could.

Competition between individuals in large, dense populations drives evolution faster. It produces more competitive types that can spread from their point of origin, displacing groups that dispersed earlier. This process can be aided by diseases, which also evolve swiftly in dense populations because transmission rates increase. Immunity builds in the dense population, but when isolated populations, not previously exposed to these diseases encounter them, they are likely to be devastated. This phenomenon of expansion from the centre is known as 'centrifugal evolution', referring to the way a centrifuge works to push things outwards; it goes a long way in explaining the demise of the Neanderthals.

The final days of the Neanderthals have been extensively researched. Until recently, it was thought that they survived on Gibraltar until about 24,000 years ago, but all such late dates are now thought to have resulted from errors. A recent study, using more rigorous methods, could not find any valid dates for Neanderthals more recent than about 39,000 years ago. It is now thought that the Neanderthals began a rapid decline starting in eastern Europe around 41,000 years ago, and that they were extinct everywhere by 39,000 years ago.[19]

It is widely believed that Neanderthals and humans overlapped briefly in Europe—for between 2500 and 5000 years. But I treat this with caution: the oldest dates for modern humans in Europe are highly questionable. The Neanderthals were the last species of *Homo* to share the planet with us modern humans. After they became extinct somewhere in Western Europe about 39,000 years ago, we were left alone. Our immediate family had been exterminated—almost certainly by our own hands. Yet this is, at best, a partial truth. Neanderthals did not die out, nor did modern humans colonise Europe.

CHAPTER 26

Bastards

The first anatomically modern humans (*Homo sapiens*) evolved in Africa around 300,000 years ago. By then, successive waves of upright apes, including *Homo erectus* and the ancestors of the Neanderthals, had been making their way into Europe from Africa for nearly two million years. Our species was destined to follow in their footsteps. By about 180,000 years ago *Homo sapiens* had pushed as far north as present-day Israel, where they may have hybridised with Neanderthals.[1] But for reasons that remain unclear, these first African expatriates did not reach Europe. It was not until about 60,000 years ago, when humans again emerged out of Africa, that our species spread.

A recent genetic study has established that the first human colonisers of Europe were a single population, derived in part from African migrants who arrived around 37,000 years ago, and who fell within the genetic variability of living Africans.[2]

Dating the chronology of hominid invasions and extinctions can lead to confusion. This is in part because the events were dated using different methods (for example, genetic comparisons and radiocarbon dates). Dates based on genetic comparisons rely on rates of genetic change, which are 'anchored' by reference to the fossil record, while

radiocarbon dates rely on estimates of decay of C^{14}. All dates are estimates, often with wide margins of error, and all methods of dating have
their own biases, which can introduce errors. We should keep in mind
that it is entirely possible that the date of Neanderthal extinction (radiocarbon dated to about 39,000 years ago) and the date for human arrival
(derived from genetic analysis as 37,000 years ago) in fact occurred in
the same millennium.

Among the oldest undisputed collection of human remains from
Europe includes partial skeletons, skulls and jaws found in the Peştera
cu Oase caves, near the Iron Gates on the Danube in Romania. The
bones have been dated to between 37,000 and 42,000 years old, with a
most likely age of 37,800 years.[3] The caves lie on a migration route into
western Europe known as the Danubian Corridor. First identified by
the archaeologist Vere Gordon Childe, many species have doubtless
followed the corridor over millions of years.

The bones found in Peştera cu Oase were first identified as those
of modern humans, but then it was noticed that they have some
Neanderthal-like features. Ancient DNA recovered from one skeleton revealed that it was a human–Neanderthal hybrid, in whom large
chunks of Neanderthal DNA (including almost all of chromosome
12) was interspersed with modern human DNA. With each generation, the DNA is mixed into ever smaller segments. The fact that the
Neanderthal DNA occurred in such large pieces in the Peştera cu
Oase individual indicates that the hybridisation event had occurred
just four to six generations back.[4] So, we know that about 38,000 years
ago, somewhere near the Iron Gates, a human and a Neanderthal had
sexual intercourse, and that the female successfully raised offspring,
which was able to reproduce.

These human–Neanderthal hybrids were probably just one of
many hybrid groups that have occurred during hominin evolution.
Evidence survives in our genes of at least one other recent event—that

between Denisovans and humans who spread east into Asia.* But what of those first-generation European human–Neanderthal hybrids? What were they like? In his epic 1903 work *The Dawn of European Civilization*, Griffith Hartwell Jones, Rector of Nuffield, uses a variety of ancient sources to reconstruct a people whom he believed inhabited Europe prior to the advent of farming. He calls them Aryans, and describes the male as follows:

> His eye was blue and fierce…he had beetling eye-brows. He was tall of stature and endowed with a powerful frame. Nurtured in a cold climate, where Nature was rugged and inhospitable, he was inured to hardship from infancy…The chase, which was his natural pastime, kept him in constant practice in the use of weapons…[5]

Written long before modern science had fleshed out our understanding of Neanderthals, it is as complete a portrait of a Neanderthal as one could want. Mix that with African genes, and the progeny would have been highly varied. Perhaps the great variation among living Europeans is an echo of the diversity seen among the first human–Neanderthal hybrids.**

In 2010 researchers announced that the entire Neanderthal genome had been sequenced.[6] No Neanderthal DNA has been found on any human Y-chromosome—the chromosome passed on only by males.[7] Unless resulting from chance, this absence could mean one of two things. It's possible that sex was only between human males

---

* The Denisovans are an extinct species or subspecies of humans known only from a few teeth and a finger bone from Denisova Cave in Siberia. They hybridised with humans, and their genes are preserved in living Asian and Australasian human populations.

** The discovery that 10,000-year-old 'Cheddar Man' had blue eyes but dark skin is to be expected in this hybrid population.

and Neanderthal females; or it may result from a curious genetic phenomenon known as Haldane's rule. Formulated by the great British evolutionary biologist J. B. S. Haldane in 1922, it states that where only one sex is sterile in a hybrid (such as in mules), it is likely to be the sex with two different sex chromosomes. In humans (and most mammals) males have an X and a Y chromosome, and females two Xs, so Haldane's rule predicts that in mammals, male hybrids are more likely to be sterile than female ones. One study hints at the possibility that Haldane's rule may have been the cause of the lack of Neanderthal DNA on the Y-chromosome of hybrids, but currently we do not definitively know.[8]

There are two main claims for fossilised human remains from Europe older than those from Peştera cu Oase. Two baby teeth, reportedly belonging to a modern human and dating to between 43,000 and 45,000 years old, were found in a cave south of Taranto in Italy, while a fragment of a human upper jaw, associated with animal bones dated to between 41,500 and 44,200 years old, comes from a Kentish cave.[9] The baby teeth have been dated by extracting material from them, but they have yielded no DNA, which means that their identification as human is based on shape alone. The Kent mandible, on the other hand, is clearly human, but its age was inferred from dates taken from animal bones preserved in the same deposit. It remains a leap of faith to assume that the human and animal bones are indeed the same age. In both cases the evidence is, I think, too thin to establish an earlier human presence in Europe.

As a palaeontologist, I am used to dealing with scraps of evidence, and resigned, courtesy of Signor-Lipps, to accepting that I'll never find the first or the last member of any species. Can we really have been fortunate enough, at Peştera cu Oase, to have discovered evidence of one of the earliest generations of European pioneers? I cannot prove it, but the site seems special—special enough, indeed, to be the one possible exception to the dictum of Signor-Lipps in this entire ecological history.

No Neanderthal bones were found at Peştera cu Oase; the hybrid's bones appear to have been washed into the caverns from outside and no rubbish dumps indicating that people inhabited the caves have been found there. We will never know for sure what happened at the Iron Gates those tens of thousands of years ago. All we can do is paint a scenario consistent with our few facts: a group of humans, on their frontier trek into new European territory, encountered a group of Neanderthals whom they ambush, killing all except the women, who are abducted and bear their abductors' children.

But there must be more to the story than this. There is something strange about the lateness of the human colonisation of Europe. As modern humans spread, one branch followed the south Asian coast and by at least 45,000 years ago had reached Australia. Europe is much closer to Africa than Australia is—so why did it take humans so much longer to colonise Europe? Part of the answer may lie in the ecological niches occupied by the early human migrants. The bands that pushed on to Australia seem to have become adept at harvesting fish and shellfish—a niche that was previously largely vacant, but which offered abundant fat and protein. Using spears, nets, stone hammers and rafts, humans could exploit the enormous bounty that existed on nearshore reefs and mudflats in a way that no other species could.

But humans living away from the coast had to compete for terrestrial resources with related species—either Neanderthals, Denisovans or *Homo erectus*—that were already adept at harvesting them. Moreover, 38,000 years ago Europe was a chilly and hostile place, in which a tropical hominid may have struggled to survive. The Neanderthals, already adapted over the millennia to Europe's harsh conditions, may have been tough competition. But then a chance event created human–Neanderthal hybrids, who quickly spread west, displacing the 'pure' Neanderthal populations.[10] It seems probable that the first human–Neanderthal hybrids possessed useful knowledge passed on by their Neanderthal mothers; and in Europe's climate, the pale skin of the Neanderthals must have been particularly advantageous as it allowed sunlight to penetrate, aiding the creation of Vitamin D.

A recent study of 50 fossils from across Europe reveals that all Europeans living between 37,000 and 14,000 years ago were descended from this founding population of human–Neanderthal hybrids. This indicates that non-hybrid humans did not make it into Europe until at most 14,000 years ago. Had scientists been around back then, they might have classified the Europeans as a new hybrid species, like the wisent. But over time, the proportion of Neanderthal DNA in the Europeans' genome decreased. In Europeans living between 37,000 and 14,000 years ago, the Neanderthal genetic inheritance averaged about six per cent. Following a migration from southwestern Asia around 14,000 years ago, this contribution was diluted to between 1.5 and 2.1 per cent (today's average). Researchers argue that many Neanderthal genes must have disadvantaged the hybrids that bore them. But just which genes, and how they acted against survival, is not clear.[11] Intriguingly, however, at least 20 per cent (and perhaps 40 per cent) of the entire Neanderthal genome survives within the genes of the European and Asian populations, because individuals have different segments of the Neanderthal genome.[12]

CHAPTER 27

The Cultural Revolution

In 1861 the French writer and artist Édouard Lartet published a drawing of a piece of bone, discovered in Chaffaud Cave in southern France, upon which the image of two hinds had been engraved. Lartet claimed that the engraving, along with other artefacts, dated from the earliest antiquity. At first his claim was met with great scepticism because European savants firmly believed that the brutish cave dwellers of the stone age were incapable of refined art. But as more pieces were discovered along with stone tools, Lartet's argument became irrefutable. Then, in 1868, paintings were discovered on the walls of a cave near Altamira in Spain, and Europeans moved closer to understanding the extent of the treasure that their most distant ancestors had bequeathed them. As more and more Palaeolithic art was discovered, it became clear that the greatest artists of Europe's stone age rivalled in vision and execution the most accomplished artists alive today.

The earliest European ice-age art is among the most striking and ingenious. One example is a magnificent 40,000-year-old half-lion, half-human carving, made of mammoth ivory and found in 1939 in Hohlenstein-Stadel, a deep cave in the Swabian Jura in Germany. The site has yielded no evidence of domestic occupation such as food remains and tools—it may have been reserved for ritual activities. The lion-person was found in more than 250 fragments.[1] Reconstructed, it stands 30 centimetres tall and has immense, almost magisterial, presence. The Swabian Jura also yielded the oldest figurative carving of a human—the Venus of Hohle Fels (Venus of Stone Hole)—which dates to between 35,000 and 40,000 years ago. Astonishingly, the world's oldest musical instrument—an ivory flute—was excavated from the Jura. It is thought to be as much as 42,000 years old, but we must keep in mind the uncertainty around such dates: the flute may be roughly contemporaneous with the Peştera cu Oase bones. These creations are attributed to the Gravettian culture, whose makers were early human–Neanderthal hybrids.

The Swabian Jura are on the Danubian Corridor, which was probably followed by the hybrid people that arose near the Iron Gates. I can imagine those pioneer beings, endowed with capacities not seen in either parent, pushing west and displacing the Neanderthals they encountered. As they settle new lands, they seek new means of expression. Neanderthal knowledge may have helped the hybrids occupy caves. In chilly Europe, caves were home for entire winters, with frozen meat and other food stored nearby. And living in caves created new imperatives and opportunities for storytelling and graphic depiction.

The flowering of artistic expression suggested by the finds from the Swabian Jura is unique in human evolutionary history. The artefacts are the oldest evidence we have from anywhere on Earth for carvings of imaginary creatures and humans, and of musical instruments. Those responsible for this artistic flowering were hybrids who, like the mule, seem to have been possessed of great reason, memory and social affection, as well as creative spirit. I find it astonishing that their novel creations were works of art, rather than the new weapons or the stone tools characteristic of earlier advances. It's as if these beings had begun a process of 'auto-domestication', with a focus on peaceful interactions rather than conflict.

It is tempting to see the sculptures and flutes as the pinnacle of ice-age cultural achievement; but these objects served a purpose, and it is the higher art that they served that we must view as the pinnacle. There is reason to believe that the art was theatre: theatre is *the* great art of Aboriginal Australia, and arguably it was the premier art in all pre-literate societies. Theatre is so important to those societies because it promotes the skills of imitation, rhetoric, expression of emotion through the whole body, and storytelling that make great hunters and leaders. Thus Shakespeare did not spring fully formed—like Athene from Zeus' head—but from a tradition that has existed at least since the creation of the first human–Neanderthal hybrids.

I can imagine those earliest performances, watched in awe by a small band in the dark of a winter night. Some may have seen the master craftsman labour over the great lion-person, and now it would come to life—in the form of a shadow cast on a cave wall. The silhouette sharpens and fades as the carving is moved before the flames of the hearth, its maker grunts in perfect imitation of the ancestor—a human-lioness in oestrus. The hybrid beast is hunting—for a human mate. Did the watchers retain a folk memory that they themselves resulted from a mating of different types: a black human and a pale Neanderthal?

From fear, apprehension and wonder, the mood changes as the sounds of the flute drift through the cavern, and the voice of the patriarch rings out. From his throat come imitations of the voices of the departed, telling of the clan's lion ancestors. Enchanted, the audience is transported to another time, another dimension. And so the long nights are spent, in the world of the first European mythology.

Having achieved their auto-domestication, the early human–Neanderthal hybrids set out to extend the achievement to the domestication of another species. One day, about 26,000 years ago, an eight-to-ten-year-old child and a canine walked together into the rear of

Chauvet Cave in what is now France. Judging from their twin tracks, which can be traced for 45 metres across the cave floor, their route took them past the magnificent art for which Chauvet Cave is famous and into the Room of Skulls—a grotto where many cave bear skulls are preserved. They walked together, companionably and deliberately, the child slipping once or twice, as well as stopping to clean a torch, in the process leaving a smear of charcoal on the cave floor. It's nice to think that the pair's Huck Finn-like exploration became the stuff of legend in their clan, for at the time Chauvet Cave's recesses had been abandoned, its art and cave bear bones already thousands of years old. And soon thereafter a landslide would seal the cave entrance. Whatever the case, the pair's adventure certainly became famous in 2016, when a large dating program of fossils and artefacts from Chauvet Cave, which included the smear of charcoal discarded by the child, confirmed that the tracks constitute the oldest unequivocal evidence of a relationship between humans and canines.[2]*

DNA studies indicate that dogs started to differentiate from wolves in Europe between 30,000 and 40,000 years ago.[3] The oldest osteological evidence is a canid skull dating to 36,000 years ago, which was found in Goyet Cave, Belgium. It is short-snouted and broad, characteristics that distinguish it from wolves—but genetic analysis places it outside the lineages of all living dogs and wolves. The skull may well have been from a group of canids that had a relationship with humans, and which subsequently became extinct. Whatever the case, Signor-Lipps warn us that children and dogs may have been associating long before this child and its canid companion strolled through Chauvet Cave 26,000 years ago.

Neanderthals and wolves had coexisted for hundreds of thousands of years—at least since the first grey wolves arrived in Europe from

---

* The footprints have been indirectly dated by radiocarbon-dating some charcoal that is presumed to have fallen from the child's torch.

Asia between 500,000 and 300,000 years ago (the oldest evidence comes from cave deposits at Lunel Viel, in France).[4] And modern humans and wolves had coexisted at least since *Homo sapiens* spread out from Africa 180,000 years ago. But it was not until the creation of the human–Neanderthal hybrids 38,000 years ago that canines and hominids began an association. One popular theory of canid domestication is that wolves began hanging around human camp-sites, hoping for scraps from kills or feeding on faeces, and that this led to a relationship. But it is more likely that domestication originated with the adoption of young animals, as still occurs today in many hunter-gatherer societies. Adoption usually occurs when a hunter kills a female accompanied by dependent young, which are brought back to camp, where they become playthings for children. In the case of wolves, it only works if the pups are no more than 10 days old, at which stage they are still in the den. If they can survive from scavenged scraps, perhaps along with breastmilk donated by a lactating mother, they may grow to adulthood. In ice-age Europe, lion and bear cubs as well as wolf pups doubtless made it into human camps as children's playthings, and, occupational health and safety not being what it is today, disasters must have occasionally befallen the adopting families. But wolves are more suitable as human companions.

A multi-decadal experiment on foxes (which are members of the dog family), carried out under the supervision of Russian geneticist Dmitry Belyayev from the 1950s onwards, has yielded important insights into the nature of the ancestral dog. Belyayev's method was simple: of the thousands of silver foxes held at a Soviet fur farm he selectively bred those that were calmer in the presence of humans. After just a few generations some foxes started to seek out human company. Breeding with these individuals resulted in foxes that showed changes in reproduction typical of domesticated animals (which often bear more than one litter per year). A few even began to wag their tails and bark—characteristics otherwise seen only in dogs. Eventually foxes were produced that had varied colour patterns, curly tails and floppy ears. A few even commenced vocalising with a sound reminiscent of human laughter. None of this was selected for—the only selection being for their level of comfort around humans. Yet, over a few decades Belyayev created foxes that behaved like domestic dogs, and were indeed suitable to keep as pets.[5]

Wolves have always had a spectrum of behaviours, from timid to aggressive, so we cannot look solely to Belyayev to explain why domestication began 37,000 years ago. I suspect that it occurred then because human–Neanderthal hybrids were the first hominids to bring puppies back to camp with the intent, not of eating them, but allowing them to become playthings.

Between the Chauvet footprints and the first widely accepted evidence of a domestic dog—the 14,000-year-old jawbone buried in a human grave in Germany—there is a wide gap.[6] The jaw indicates the beginning of a long tradition of the interment of dogs with people, which reveals a deep attachment between some people and canids. By 4000 years ago the first domestic breeds (which were greyhound-like) had emerged, and dogs were on the way to becoming the highly modified creatures that many of us live with today. It's as if the human–Neanderthal hybrids of 38,000–14,000 years ago were happy to coexist with wolf-like dogs, while later peoples preferred more modified types of canine companions.

It has recently been proposed that a second group of wolves were domesticated independently in China or in southeastern Asia.[7] Several factors make this theory difficult to test, one being that genetics provide no clear guide because no living dogs are more closely related genetically to the wolves of any particular region than any other dogs, probably because of a repeated mixing of dog and wolf genes. And in the millennia since the first domestication, breed selection has further scrambled the dog genome, making it difficult to pinpoint geographic origins. The archaeological record is of only modest assistance in clarifying things: we have fossils of dogs dating back 36,000 years in Europe, 12,500 years in east Asia, but only 8000 years in central Asia. The 4500-year-long difference in dates might be evidence that dogs did not reach east Asia from Europe and were domesticated independently there. But Signor-Lipps may also have something to say about that.

CHAPTER 28

Of Assemblages and Elephants

When humans arrived in Europe, an already chilled Earth was becoming more intensely cold. The substantial ice caps had lowered sea levels by 80 metres below today's level. In the millennia thereafter, the glaciers would wax so thick, and sprawl so far, that sea levels would drop by a further 40 metres. As a result, there was no Baltic Sea, and you could have walked from Norway to Ireland, even if it meant crossing ice and a few rivers. The cooling was fast by geological standards, but it would have been imperceptible to anyone living, being at least 30 times slower than the warming trend that we are currently experiencing courtesy of greenhouse-gas pollution. It was, nonetheless, forcing changes in the abundance and distribution of flora and fauna across Europe.

After the glacial advance that occurred around half a million years ago, many European animals came to exist as two related or ecologically similar types—one of which dominates in the cold phases, and the other during the warm periods that have prevailed just 10 per cent of the time over the past million years. The woolly mammoth and Europe's straight-tusked elephant are such a pair, as are the woolly rhino and Europe's extinct forest rhinos. The carnivores were not as likely to split as the herbivores because they were better able to cope with a variety of climates by sheltering in caves. The spotted hyena, for example, was once distributed from the edge of Europe's polar desert to equatorial Africa.

Europe's mammals are described by scientists as comprising faunal assemblages—groups of species that typically occur together. Let's look at five large creatures from Europe's ice-age, warmth-loving faunal assemblage: the straight-tusked elephant, two rhinoceroses, the hippopotamus, and a water buffalo. The largest of these was the straight-tusked elephant, which first reached Europe from Africa around 800,000 years ago. They could grow to be very large indeed; one male is estimated to have weighed 15 tonnes, which is half as large again as the biggest elephant living today.

Europe's straight-tusked elephants probably had a herd structure similar to that of other elephants, in which females and young live in small groups, while the larger males were either solitary or congregated in bachelor herds. Straight-tusked elephants could be found in forest and more open habitats, including the warmth-loving oak forests and varied vegetation types that continue to grow around the Mediterranean and in southern and central parts of Europe today. It's reasonable to suppose that, were they still around and left unmolested by hunters, Europe's straight-tusked elephants would thrive in forests from Germany to Sicily, and from Portugal to the shores of the Caspian Sea.

The European straight-tusked elephant was long classified in an extinct genus, *Palaeoloxodon*, the various species of which could once be found from western Europe to Japan and east Africa. But in September 2016 researchers announced that they had successfully extracted DNA from the bones of a 120,000-year-old straight-tusked elephant from Germany and identified its nearest relative, *Loxodonta cyclotis*—the African forest elephant.[1] Africa has two elephant species— the rainforest-dwelling type, and the more familiar and widespread savannah elephant—which split between five and seven million years ago. When the full research findings of this work were published in February 2018 the story got even more astonishing. Genes from an ancestor to both African elephant species comprise the largest element in the European straight-tusked elephant genome, with the next largest contribution (between 35 and 39 per cent) coming from the African forest elephant, and much smaller contributions from both the woolly mammoth and African elephant. The European straight-tusked elephant is thus a complex hybrid.[2]

Pulling all the data together, it seems likely that the European straight-tusked elephant arose in Africa before today's living African species separated. Then, at some time before 800,000 years ago, it hybridised extensively with the African forest elephant. Finally, limited interbreeding with both the woolly mammoth and the African elephant occurred. Just how the taxonomists will classify such a creature is yet to be resolved.[*]

Both European straight-tusked elephants and African forest elephants have long straight tusks; those of older African forest males almost touch the ground. This contrasts with the curved tusks of Asian and other African elephants, and mammoths. The straight-tusked elephants include both the largest and smallest of all elephants. The largest living African forest elephants can reach six tonnes in weight, but the 'pygmy' elephant living in the Congo averages just 900 kilograms when adult. European straight-tusked elephants could weigh as much as 15 tonnes, but some island-dwelling forms were pig-sized.

It seems almost unbelievable, but until 2010, scientists did not know that the two living African elephants were distinct species. But 110 years earlier, one of the most eccentric zoologists of all time, Paul Matschie, had identified the African straight-tusks as different. Matschie started

---

[*] Also known as the African forest elephant.

his career as a volunteer at the Berlin Zoological Gardens, and, despite his lack of formal qualifications, in 1895 was appointed its curator of mammals. In the habit of wearing pince-nez glasses and sporting a splendid moustache, by 1924 Matschie had become that august institution's director.

Over the years of working with zoo animals, Matschie developed his own highly unusual theory of classification. Known as 'the theory of the half-sided bastards', it declared that each major watershed on Earth harbours a distinct species of any given kind of animal. If the animals living in the watersheds ever met on the ridges dividing them, the creatures might hybridise. Such hybrids could be recognised as 'half-sided bastards', because they would resemble one parent on one side of their head, and the other parent on the other.

I can imagine Matschie's underlings coming to 'Herr Director' with the odd goat that had one horn straighter than the other, or a deer with one antler more elaborate than the other, or indeed an elephant with one tusk straighter than the other, hoping to curry favour. Such novelties may well have encouraged Matschie until his bizarre theory became an unshakeable foundation stone of his thinking. Indeed, on the joyous occasion that an unusual skull with asymmetrical horns or tusks turned up, Matschie celebrated by describing *two* new species based on the one specimen—one for each supposed unknown parent species, which he reasoned must still be lurking in their unexplored catchments. It is easy to understand why much of Matschie's work was ignored. Yet who would have believed that, when it came to straight-tusked elephants, the truth was even more fantastical?

Perhaps one day Europeans will decide to return elephants to their continent. If so, they would be well served by starting with forest elephants from Africa. But they should not wait too long as the beasts are becoming increasing endangered. Part of the problem is their slow rate of reproduction. Straight-tusked elephants take about 23 years to reach sexual maturity, and thereafter give birth only once every five or six years. The African savannah elephant, in contrast, matures at about twelve years and can give birth every three to four years. The slower the reproductive rate, the more impact hunting has. Between 2002 and 2013, 65 per cent of the African forest elephant population was killed, mostly by poachers seeking ivory. At that rate, extinction will occur in the next few decades.

Many people find the prospect of elephants wandering the forests of Europe ridiculous, or even dangerous. Yet they accept that Africans must share their homes with the ponderous creatures. I think that we should take the long view and share the burden of conservation more equally. But bureaucracies keep getting in the way. The IUCN (International Union for the Conservation of Nature), for example, restricts use of the word 'reintroduction' to species that have become extinct locally or Europe-wide no more than 200 to 300 years ago. Just why this is I cannot imagine, but I urge the IUCN to appoint more palaeontologists to its committees!

What drove Europe's straight-tusked elephants to extinction? We can never be certain, but we can look at patterns of climate, predation and distribution. Fossils reveal that as the ice-age gripped the continent, straight-tusked elephants retreated to the warmer southern peninsulas of Spain, Italy and Greece. This would have limited their overall population size and divided it into sub-populations that could not easily intermix, making them more vulnerable to extinction. Straight-tusked elephants doubtless had their predators too, with lions and spotted hyenas taking the odd calf. There's also good evidence that the Neanderthals hunted them. A 400,000-year-old straight-tusked elephant skeleton found in the Ebbsfleet Valley near Swanscombe in Kent was surrounded by stone tools indicating that it had been butchered, while marks on the bones of a second individual found in Britain suggest that it had been cut up. In both cases, however, it's possible that the Neanderthals were scavenging from a carcass. But a third skeleton, found near Lehringen, Germany, was found lying on a 125,000-year-old wooden stabbing lance, which appears to have been used to kill it.[3] Several other elephant skeletons have been found alongside stone tools in Spain, Italy and Germany, so it seems safe to say that Neanderthals could kill adult straight-tusked elephants.

Fossils suggest that straight-tusked elephants made their last stand on the European mainland in Spain, about 50,000 years ago. This is curious: 50,000 years ago, the ice had not yet fully extended and substantial areas of forest persisted. Indeed, conditions were not greatly different from those of previous glacial advances that the elephants had endured. Could it be that straight-tusked elephants survived longer on mainland Europe? It is a question that Signor-Lipps have a firm opinion on. And, indeed, a single 37,000-year-old image of a furless elephant from Chauvet Cave, France, may depict this species.

Europe's straight-tusked elephants survived on various islands in the Mediterranean for thousands of years after they vanished from mainland Europe. All of the island populations were dwarfs, some being very tiny indeed. Cyprus's straight-tusked elephants, for example, were only a metre high at the shoulder and weighed a mere 200 kilograms. These tiny elephants survived until about 11,000 years ago, and they shared the island with the smallest hippo known, *Phanourios minor,* which was the size of a sheep. Cyprus was settled by humans at least 10,500 years ago, and the campsites of these early Cypriots have been discovered in caves at Aetokremnos (Vulture's Cliff) on the Akrotiri Peninsula.[4] The bones of hippos are found in layers immediately below the human camps, but it is not known with certainty whether humans hunted the hippos, or indeed the elephants. The island of Tilos in the Dodecanese may have offered a last refuge. Its elephants, which averaged two metres high at the shoulder, survived until about 6000 years ago. This date, however, deserves further investigation, for

Tilos supported a population of humans for thousands of years before that—and, at least on small islands, the archaeological evidence from elsewhere suggests that humans and elephants don't coexist.

If a changing climate was to blame for the extinction of the straight-tusked elephants, why should they have survived on islands long after those on the adjacent mainland vanished? Surely climatic shifts would affect islands and the adjacent mainland equally? The fact that the pygmy elephants survived on Cyprus until around the time that humans discovered their island is, I think, telling. The cold phases of the ice age were bad news for straight-tusked elephants, but there was another, more decisive influence at work—humans.

CHAPTER 29

Other Temperate Giants

After elephants, the largest creatures humans encountered in Europe were rhinos. Merck's rhino and the narrow-nosed rhino were close relatives, having diverged from a common ancestor a million or so years ago. The larger Merck's rhino (which could weigh up to three tonnes) was a specialised browser, much like Africa's black rhino, while the narrow-nosed rhino fed on grass, as does Africa's white rhino. Although ecologically similar, neither species was closely related to the African rhinos living today.* Instead, rather surprisingly, Europe's extinct rhinos were related to the Sumatran rhinoceros, which is critically endangered. The range of Europe's rhinos extended far to the east: Merck's rhino as far as Afghanistan, and the narrow-nosed rhino as far as eastern China.[1]

Despite the abundance of their fossilised remains, both species remain under-researched. A study of DNA could reveal much about their evolutionary relationships, and a careful dating program may tell us more about their extinction. They seem to have survived in Spain (Merck's rhino) and Italy (narrow-nosed rhino) until about 50,000 years ago. But some 37,000-year-old depictions in Chauvet Cave, France, may

---

* Africa's rhinos diverged from other rhinos around 24 million years ago.

also represent one of these species. The Chauvet depictions show beasts with dark bands around their girths, raising the intriguing possibility that they were patterned like Holstein cattle.

Hippo remains dating to around 100,000 years ago have been found in sediments in the lower reaches of the Thames and in the Rhine and Danube rivers. Hippos do not like severe frosts, and they retracted southwards as the climate cooled, before disappearing altogether from Europe long before humans arrived. The final member of temperate Europe's big five was the water buffalo. Its fossilised remains abound in the river valleys of western and central Europe, particularly in the Netherlands and Germany. There appear to have been minor differences in horn shape between the European fossils and the living Asian water buffalo, prompting some to place the European fossils in their own species. Whatever the case, the extinct European population was very similar to the living Asian river buffalo. Its genetics and the date of its extinction have not been adequately researched, though the species may have survived in eastern Austria until about 10,000 years ago.[2]

Water buffalos are such useful animals that they were reintroduced to Europe. The Lombard king Agilulf may have been the first, bringing them to the Milan area in 600 CE. Armenia also received them early, and introductions have continued to the present, with domestic populations thriving today across Europe, from Romania to the United Kingdom. But perhaps the best place to see them is on the plains around Salerno in southern Italy, where their milk is used to produce the delicious mozzarella that the region is famous for.

Three of temperate Europe's 'big five' are not extinct or have surviving close relatives: the straight-tusked elephant, the hippo and the water buffalo. It's just that none has survived continuously in Europe, though the water buffalo was reintroduced early on and exists in domestic form. But other mega-herbivores flourished in temperate

Europe before the arrival of *Homo sapiens*. In descending order of size they were: the aurochs, giant deer, cave bear, red deer, wild boar, fallow deer and roe deer. Of these, only the cave bear and giant deer are extinct, while all the rest survive in Europe in one sense or another.

Cave bears and European brown bears are close relatives, but brown bears occur in Europe, Asia and North America (where they are known as grizzlies) while cave bears were restricted to Europe. Both are descended from the ancestral Etruscan bear that existed a little over a million years ago. Brown bears and cave bears coexisted but seem to have divided the ecological niche by size and diet. European brown bears are today largely herbivorous, but bone analysis shows that in the past they ate lots of meat. Cave bears, in contrast, were purely herbivorous.[3]

Cave bears probably looked like oversized brown bears with dished foreheads. At a tonne in weight, they were twice the size of the largest European brown bears, and had skulls up to three-quarters of a metre long. The cave bear population started to decline about 50,000 years ago, contracting westwards until the last known populations remained in the Alps and adjacent areas, where they became extinct about 28,000 years ago.[4]* Both Neanderthals and human–Neanderthal hybrids hunted them: a 29,000-year-old vertebra found in Hohle Fels cave in the Swabian Jura retains the flint spear head that killed the animal, and cave bear bones (mostly juvenile) from the site have the marks of butchering and skinning. The evidence from Hohle Fels indicates that cave bears were important prey for the hybrid human–Neanderthals living there: for at least 5000 years they consumed their flesh, used their skin for rugs or clothing, their teeth for ornaments, and burned their bones for warmth.[5]

Countless fossils of the giant deer have been unearthed from Ireland's peat bogs, and it seems that in the nineteenth century no

---

* So far, an extensive dating program for late-surviving cave bears has only been conducted in the Alps. It's possible that a few lingered on elsewhere in western Europe.

Irish baronial manor was complete without the skull of an 'Irish elk' in the entrance hall. Its fossils occur across a vast swathe of Eurasia, from Ireland all the way to China. At more than 600 kilograms in weight the giant deer was the size of a moose. Its enormous antlers could weigh 40 kilograms and measure more than 3.5 metres tip to tip. Cave paintings indicate that it was pale in colour with a dark stripe over its shoulders. Its nearest living relative is the much smaller fallow deer, whose antlers are similar in shape.

Traditional explanations for the extinction of the giant deer focus on its antlers as being somehow maladapted to altered vegetation or climatic conditions, or on a decrease in accessible nutrition. But the evidence we have does not fit either theory. The last records are from northern Siberia, where it survived until about 7700 years ago, at which time the climate was broadly similar to today's. Moreover, the most recent fossils show no evidence of malnutrition. Two skeletons found on the Isle of Man have been dated to about 9000 years ago.[6] By this time, the Isle of Man had been cut off from the rest of Britain by rising seas for 3000 years. Both skeletons are from much smaller individuals than those that lived in the region just a few thousand years earlier. Their small size might be the result of island living, or perhaps a warming climate. Perhaps these 'dwarves' survived on the Isle of Man because their home had not yet been invaded by people?

Ice-age Europe's rich and varied large carnivore fauna comprised brown bears, lions, spotted hyenas, leopards and wolves. Of these, only the brown bear and wolf survive in Europe today. The cave lion, an enormous predator about 10 per cent heavier than today's lions, is only modestly distinct from the surviving lion species, having diverged about 700,000 years ago. Its appearance is well known from cave art, ivory carvings and clay figurines: it lacked a mane, was the same colour or a little lighter than the modern lion and had the same ears and tufted tails. But it had a dense underfur, and some may have been faintly striped.[7*] It had one of the widest distributions of any mammal, being found from Europe to Alaska, and ranging far into the freezing north. A pair of week-old cubs, at least 10,000 years old, was recently discovered, preserved in permafrost, in Siberia.

The diet of cave lions seems to have varied regionally. Some specialised, preying on reindeer, while others preferred young cave bears.[8] After the arrival of humans in Europe, cave lions began to decrease in size. The most recent individual known, from northern Spain, was no larger than a living African lion. The discovery of a 'living floor' in the lower gallery of La Garma Cave, near Cantabria in Spain, which had remained undisturbed for 14,000 years, provides a remarkable insight into interactions between humans and the last of the cave lions.[**] Inside the cave, whose walls are decorated with art, were the ruins of three stone huts, located about 130 metres from the original entrance and dating to between 14,300 and 14,000 years ago. They appear to be the result of a single, relatively brief occupation that was terminated when a rockfall sealed the chamber. The bones of horses, aurochs, red deer, reindeer, brown bear, fox and spotted hyena are clearly the leftovers from meals. But around one of the huts lay nine claw bones of a cave lion. They were cut in a way that indicated that the creature had been skinned. Researchers believe that the claws, which are from the front paws, formed part of a lion-skin rug in one of the huts.[9] Deposits of bones show that the hunting of carnivores by people had increased by the time La Garma Cave was occupied. Were carnivores being targeted because large game was scarce? Or were developments in hunting technology making it easier to kill lions and hyenas? Whatever the case, the La Garma claws are the last evidence of the cave lion in Europe.

---

[*] The discovery of fur preserved in the Siberian permafrost has revealed details of colour and underfur.

[**] A living floor is, in archaeological terms, the floor of a cave upon which people lived and which retains evidence of their activities.

After the cave lion, the next largest predator was the cave hyena. Attaining more than 100 kilograms in weight and at least 10 per cent larger than the spotted hyenas of Africa today, it was a formidable predator capable of killing a woolly rhinoceros. Despite its large size, genetic studies show that it belonged to the same species as today's African spotted hyena.[10] The species first arrived in Europe about 300,000 years ago, about the time of the extinction of the giant hyena, *Pachycrocuta*, which was twice its size. The cave hyena was widespread and abundant in Europe and north Asia, ranging from Spain to Siberia. Although present in virtually all habitats, it preferred to den in caves, so its distribution, especially in cold, northern areas, may have been limited to limestone and other rocky regions where caves form. Neanderthals and hyenas probably competed for caves, and hyenas seem to have occasionally pilfered Neanderthal kills, while Neanderthals occasionally killed and ate hyenas.

A study of climatic variability and hyena distribution in Europe indicates that the species' extinction cannot be blamed on a shifting climate. In fact, the changing climate of Africa (where it survived) seems to have been even more challenging to the hyenas.[11] While it is tempting to cite the arrival of humans as the cause, evidence is sadly lacking. All we know for sure is that 20,000 years ago, cave hyenas began to disappear from Europe.

Leopards, living males of which weigh between 60 and 90 kilograms, and females between 35 and 40 kilograms, were the next largest of Europe's vanished predators. They once occurred as far north as England and might have survived until 10,000 years ago in western Europe.[12] Today, Europe's last leopards maintain a claw-hold in Turkey and Armenia, where they are critically endangered, with perhaps only a few dozen surviving. But leopards do not give up easily. In the 1870s, one swam 1.5 kilometres from Turkey to the Greek island of Samos. The creature was trapped in a cave by a local farmer and eventually killed, but not before it had inflicted fatal wounds on its persecutor.

CHAPTER 30

Ice Beasts

When we hear the words 'ice age', we think of those frigid, treeless regions that at times expanded to become the largest habitat on Earth. The frozen north also had its big five, which included those iconic species, the woolly mammoth and the woolly rhinoceros. Both species were named in 1799 by Johann Friedrich Blumenbach, who is perhaps most famous for his naming of the races of humans. He believed that everyone was descended from Adam and Eve, and that the differences between the races resulted from environmental factors active since people had dispersed from the Garden of Eden, which was thought to have been in the Caucasus. Blumenbach believed that, given the right conditions, people would revert to their original Caucasian form. He possessed the skull of a Georgian woman which he thought was close in form to Eve's. He probably thought of his fossil mammoth and rhino as close to the God-created individuals that inhabited Eden, for he named the woolly mammoth *primigenius* (meaning 'first') and the rhino *antiquitatis* ('of the good old days'). Essentially, Blumenbach's classification relied on archetypes—the ideal of species as they were at the Creation.

The ice age was nearly two million years old by the time the woolly mammoth evolved. The Anglian glaciation, from 478,000 to 424,000 years ago, was particularly cold, and it marks a time of momentous change. One such change, which resonates even today, was a topographic alteration that has recently been termed the 'geological Brexit'. Prior to the Anglian glaciation, a high chalk ridge had run from what is now the cliffs of Dover all the way to Calais. During warm phases, when the seas rose, this ridge provided the only dry land corridor linking Europe to peninsular Britain, and it would have acted as an ice-age highway for all land creatures migrating east or west.

By about 450,000 years ago melting glaciers had created a gigantic lake to the north of the chalk ridge, which filled until water began to pour over in a series of cascades so immense that they created 'plunge pools' up to 140 metres deep at their bases.[1] A second breaching, about 160,000 years ago, completed the destruction of the ancient land bridge. During warm times, Britain became an island. The only land route was during cold phases, when sea levels fell, favouring colonisation by cold-adapted land creatures.

The Anglian glaciation acted as a spur for the development of a unique mammal assemblage known as the mammoth steppe fauna. This fauna, which would come to dominate Europe during cold phases, first occurred about 460,000 years ago.[2] Its 'core fauna' consisted of the woolly mammoth, woolly rhino, saiga, muskox and the arctic fox.* All except the woolly rhino evolved in the northern Arctic, and all had been evolving for several million years before their modern forms appeared.

The woolly mammoth is the defining species of ice-age Europe in the sense that it is credited with helping create and maintain the largest habitat ever to exist on land—the mammoth steppe. Alaskan palaeontologist R. Dale Guthrie coined the term 'mammoth steppe'. His interest in the vanished habitat was piqued by the observation that

---

* A core fauna denotes a group of species that are always found in association.

some of the regions the mammoth once roamed are today poor habitats, consisting of a thin layer of boggy vegetation lying over permafrost
in which the nutrients are locked. These regions are barely capable of
supporting bison, much less mammoths. He theorised that the very
different habitat that existed during the ice age—was one created by
the actions of the mammoths themselves. He thinks that mammoths,
whose tusks acted as huge snowploughs (and are often found to be
worn flat on the underside from such use), uncovered grass on which
many beasts fed, so that by spring the vegetation had been cut back to
bare earth, allowing the sun to warm the soil. This promoted swift
new growth and prevented the build-up of boggy vegetation that
could freeze into permafrost, locking away nutrients. In effect, intense
grazing created a hugely productive habitat.

While supremely important in terms of ice-age ecology, the woolly
mammoth itself is somewhat inflated in the public imagination. Some
mammoths, including America's Columbian mammoth, were indeed
among the largest elephants ever, but woolly mammoths were on
average no larger than Asian elephants. Asian elephants and woolly
mammoths are close relatives, their ancestors having diverged in Africa
only four to six million years ago.[3] It was not until about 800,000 years
ago that the classic woolly mammoth first appeared in Siberia. By half
a million years ago it had reached western Europe.[4]

Quite apart from their luxurious covering of long hair and fur,
woolly mammoths looked very different from today's elephants, having
a high-domed head, a pronounced shoulder hump of fat, and a back
that sloped steeply to the rear. Their ears were tiny, their tusks were so
curved that some crossed, their tails were short, and they were equipped
with a 'clapper valve' that could cover their anus to protect it from cold.

Cave art has so exquisitely captured these majestic creatures that,
seeing the images, we immediately comprehend the great, shaggy,
hump-shouldered beasts, looming out of the cave wall and travelling

in single file into the blizzard. Carcasses preserved in permafrost enable us to touch their long fur, study their parasites and prise lumps of food from between their teeth. It is even said that Siberian explorers feasted on mammoth flesh preserved in the permafrost.* More recently, advances in forensic DNA techniques have allowed us to recover the mammoth's entire genome.

Following its arrival in the fossil record, the woolly mammoth appears across the breadth of Europe whenever the ice advances, except in the temperate refuges of the south. Yet, by 20,000 years ago it was in trouble. Detailed studies of mitochondrial DNA show that, beginning about 66,000 years ago, North American mammoths colonised Eurasia and gradually replaced existing types of mammoth until the Eurasian mammoths became extinct about 34,000 years ago. Strangely, the North American migrants don't show up in western Europe until 32,000 years ago, leaving a 'mammoth gap' of 2000 years. Between 21,000 and 19,000 years ago, woolly mammoths are again absent from central continental Europe, and by 20,000 years ago they had gone from Iberia. They return briefly to Germany and France about 15,000 years ago, recolonising as far west as Britain, but within a millennium they vanish again. An adult male mammoth and four juveniles, trapped about 14,500 years ago in a boggy 'kettle hole' left by a retreating glacier in Shropshire, are the most recent record in the UK. With the demise of the last German mammoths about 14,000 years ago, the beast is gone permanently from Western Europe.[5]

Eurasia is far larger than North America, and it was always home to the largest section of the mammoth steppe. As Darwin's rule informs us, creatures from larger regions more often invade smaller areas, so it

---

* There appear to be no authenticated instances of modern humans eating mammoths. The infamous account of diners at New York's Explorer's Club feasting on a 250,000-year-old Alaskan mammoth in 1951 never occurred.

seems anomalous that North America's mammoths replaced Eurasian types, rather than the other way around. But this assumes equal population densities: North America did not have mammoth-killing upright apes, so it's possible that the mammoth population of North America was denser than that of Eurasia.

The last European mammoths survived on the Russian plain, including the region that is now Estonia, until about 10,000 years ago. Incidentally, the remains of the last known mammoth in Europe were discovered in grim circumstances. In 1943, during World War II, desperately cold and hungry Russians dug into a peat bog near Cherepovets, 500 kilometres west of Leningrad, searching for fuel to keep warm. They found little peat, but at a depth of two metres they encountered huge bones, that were found to be the remains of a single mammoth. Someone took the time to deposit the bones in the local museum, and in 2001 some rib fragments were radiocarbon dated, placing the mammoth at between 9,760 and 9,840 years old.[6]

The range of the mammoth was contracting swiftly by 20,000 years ago, yet the great warming and ice melt did not begin until about 7000 years after that, so the pattern of mammoth decline is not a perfect match for climatic change. But humans had begun colonising the mammoth steppe, pushing as far north as the Arctic Ocean, perhaps accompanied by the domestic version of that tundra veteran, the wolf. It seems possible that by 15,000 years ago almost all the mammoth habitat on the Eurasian mainland was accessible to human hunters, and that only lonely Wrangel Island in the Arctic Sea lay beyond their grasp. It was there that the mammoth survived—for a full 6000 years after their extinction on the mainland. Wrangel lies 140 kilometres north of the Siberian mainland and is 7600 square kilometres in extent. Its mammoths were island dwarfs. The earliest human presence detected on Wrangel dates to about 3700 years ago, and the most recent Wrangel mammoth found date to about 4000 years ago, so (with Signor-Lipps and the limited precision of dating in mind) the arrival of humans is most likely the cause of their extinction.[*]

The extinction of the woolly mammoth, according to some researchers, sounded the death knell for the mammoth steppe, an ecosystem dominated by nutritious grasses, herbs and willow shrubs that thrived in a cold, dry climate. Bounded by great ice sheets that isolated it from the sea, it was a dry, dusty place of clear skies in which spring warmth could quickly penetrate the soil and trigger a vigorous growing season that provided abundant food and allowed giant mammals to flourish. About 12,000 years ago the mammoth steppe went into rapid eclipse. The Altai-Sayan region in Mongolia supports a last relic. It is the only region where saiga antelope and reindeer—two core mammoth steppe species—coexist today. In the absence of mammoth, climatic stability may have permitted this remnant to survive.

The mammoth steppe and other northern habitats supported a wide variety of mammals in addition to mammoths, including the woolly rhino, bison, horse, moose, muskox, reindeer, saiga antelope and Arctic fox. All are familiar as living creatures—except the woolly rhino. This member of the rhinoceros family originated not in Siberia, but on the Tibetan plateau. Its nearest living relative is the Sumatran rhino, from which it separated about four million years ago. At 1000 kilograms in weight, the Sumatran rhino is the smallest living rhino species, and today it survives only in tropical rainforest. But a more northern subspecies exists in Burma, which is larger and has hairy ears.[**] Perhaps four

---

[*] Mammoths also survived on St Paul Island, Alaska, until about 5000 years ago.

[**] The subspecies, which had a very long second horn, and was much larger than the Sumatran animals, is known as *Dicerorhinus sumatrensis lasiotis*. Although the last confirmed specimens date to the nineteenth century, rumours suggest that it may still exist. It would be interesting to compare its DNA with that of the woolly rhino.

million years ago, something like it wandered into ever higher elevations in the Himalayas, giving rise by 3.6 million years ago to an ancestral woolly rhino. As the ice ages set in, the woolly rhinos found congenial conditions in the mammoth steppe that took hold across Eurasia, and they spread from France to eastern Siberia.

Two complete woolly rhinos were found preserved in tar seeps near Starunýa in Ukraine in 1929. These, along with mummified pieces preserved in permafrost, have enabled us to reconstruct a great deal about the vanished creatures' appearance and lifestyle. Like the woolly mammoth, the woolly rhino was not as large as legend suggests. Weights have been estimated for females only; they reached about 1500 kilograms. Males are likely to have been larger but did not weigh as much as Africa's white rhino. The woolly rhino had a broad upper lip like the white rhino, perfectly adapted for cropping a sward consisting of meadow plants, grasses and herbs.

Most of the woolly rhino's anatomical peculiarities involve adaptation to life in the frigid north. Its covering of dense wool and long hair, short tail and short, narrow, leaf-shaped ears (unlike the more rounded ears of living rhinos), all limit heat loss. Its two horns were flattened in such a way that, had you seen it from straight on, they would have appeared very narrow. Wear reveals that they were used as snow-sweeps as the creature moved its head from side to side.[7] Woolly rhinos appear to have become extinct in Britain by about 35,000 years ago, with the last inhabiting Scotland.[8] They may have survived in western Siberia until 8000 years ago.

To fill out this ice-age herbivore bestiary, it remains to meet a couple of astonishing creatures that our ancestors may have encountered in Europe. The 'unicorn beast' (*Elasmotherium sibericum*) was a kind of long-legged rhino that weighed 3.5–4.5 tonnes—as much as an elephant. The very largest individuals inhabited the Caucasus region, on the border of Europe and Asia. The unicorn beasts were runners and grazers. Their popular name derives from the fact that they had a single horn which, judging by the indent it left on the skull, was a metre in circumference at the base and two metres long. A wound to a knee bone suggests that the great creatures used their horns to joust—most likely in altercations over females. Recently discovered fossils indicate that unicorn beasts survived to 29,000 years ago in the Pavlodar region of Kazakhstan.[9] A rough outline of a hump-shouldered, single-horned creature drawn on the wall of Rouffignac Cave in France may be evidence that their range once extended to western Europe.

On 16 March 2000, the Dutch fishing trawler *UK33* plucked the jawbone of a strange creature from the depths of the North Sea at Brown Bank, off the Norfolk coast. After about six weeks in the hands of fishermen, during which time it lost all but two of its teeth, the fossil came into the possession of Dutch palaeontologist Klaas Post, who recognised it as the right lower jaw of the scimitar-toothed cat, *Homotherium*. At up to 440 kilograms in weight, *Homotherium* was far larger than a lion, and its diet matched its size. A lair discovered in Friesenhahn Cave, Texas, was filled with the bones of juvenile mammoths. When the jaw was radiocarbon dated, it proved to be just 28,000 years old.[10] Before this discovery, the species was presumed to have become extinct in Europe about 300,000 years ago. Signor-Lipps would have been pleased!

You might think that, in a contest between such a beast and a human, the result would be a foregone conclusion. But the history of the scimitar- and sabre-toothed cats suggests otherwise. These cats evolved in Africa, but by 1.5 million years ago, *Homotherium* was extinct there, and the sabre-tooths by one million years ago. *Homo erectus* evolved in Africa around two million years ago, and by a million years ago its brain size had increased and its technology improved.

Both scimitar- and sabre-toothed cats survived longer in Europe—until about half a million years ago, by which time the ancestors of the Neanderthals had arrived. Highly efficient hunters who used fire, the Neanderthals may have outcompeted both sabre-toothed and scimitar-toothed cats. Both types however, continued to thrive in the Americas until about the time humans arrived 13,000 years ago.[11] This global history of extinction suggests that these great cats began to decline whenever humans or their ancestors showed up.

The discovery of the 28,000-year-old *Homotherium* bone should not be taken as evidence that humans and scimitar-toothed cats overlapped for long in Europe. *Homotherium* was gone from the more temperate areas of Europe by about half a million years ago, and the fossil dates from an extremely cold period. The great cat may have survived only in the far north, which was largely beyond the range of human settlement until about 15,000 years ago.

CHAPTER 31

What the Ancestors Drew

Great treasure troves of art, preserved in caves sealed for millennia by rockfalls, have been discovered in Europe, providing a glimpse of a lost world of European creativity. Arguably, the finest of this art is the oldest—from Chauvet Cave in southern France.* But if we are to see the world through the eyes of the mammoth hunters we must look at ice-age art as a whole. And there is no better guide for this than Alaskan hunter, artist, palaeontologist and naturalist R. Dale Guthrie—the man who named the mammoth steppe.

In his book *The Nature of Paleolithic Art*, Guthrie makes the point that ice-age art focuses on a particular subset of subjects. There are no depictions of buttercups, babies or butterflies, despite the fact that all must have abounded during the ice age. There are, indeed, almost no depictions of plants. The main focus of ice-age art, as far as food is concerned, is large mammals, with a lesser focus on edible birds, fish and insects, although almost all of the insect images represent the larvae of warble flies, a kind of maggot that lives under the skin of reindeer, and which is a delicacy among Arctic people today.[1]

---

* The oldest paintings in the cave are about 33,000 years old, but the oldest habitation levels date to 37,000 years ago. More work is required before a full understanding of the chronology of the site is clear.

Guthrie also observed that the ice-age artists didn't depict generalised animals, but creatures of a particular sex and age, behaving in typical ways. For example, reindeer are depicted as either male or female (easily distinguished by their antlers), and in pre-rut (fat) or post-rut (thin) condition. Finally, he explains that the great majority of ice-age art is the work of 'learners' whose sketches and drawings contain many mistakes or are merely casual attempts.

The three great Palaeolithic art galleries of Europe are the work of master painters: Chauvet Cave in southern France, dating from 37,000 to 28,000 years ago; Lascaux Cave, also in southern France and dating to 17,000 years ago; and Altamira Cave in northern Spain, dating to between 18,500 and 14,000 years ago (though some of its images could be 36,000 years old).[2] Although spanning a possible 25,000 years, and despite each having its peculiarities, the art in these galleries shares common elements of style, purpose and subject matter.

The images were drawn using similar materials, among the most important of which were ochre, haematite and charcoal. The most frequently occurring subjects are aurochs, bison, horses and deer. Chauvet's depictions can be identified as 13 species, including a variety of carnivores, such as lions, leopards, bears, and cave hyenas. A thin, hairless elephant (possibly a straight-tusked elephant) is depicted, as are rhinos. The Chauvet rhinos appear to be hairless, often with a dark belt around their girths. All other ice-age depictions of elephants appear to be woolly mammoths, and other depictions of rhinos show more uniformly coloured beasts with shaggy fur—almost certainly woolly rhinos.

Lascaux has by far the greatest abundance of art, with about 2000 images, including a single human. Curiously, reindeer, the principal food of the inhabitants of Lascaux to judge from the bones preserved in the cave, is represented by just one image. Altamira, the most recent, has the fewest images. It contains a depiction that is possibly of a boar

(Guthrie identifies it as a poorly executed bison). It is striking just how many species of Europe's woodlands (including aurochs, deer and possibly woodland rhinos) are depicted at these sites.

Scenes of animals defecating are common in ice-age art, leading some experts to suspect that a 'cult of defecation' existed among our ancestors. Guthrie, however, argues that many large mammals defecate before fleeing, so we are seeing depictions of animals at the beginning of a chase. Other animals show spears sticking out of their bodies, or with guts hanging out of a belly wound, or coughing up what appears to be lung blood, denoting that the animal is dying. Another feature of the art is an abundance of red spots, which Guthrie interprets as blood droplets—the spoor that a wounded animal leaves as it flees. The case can thus be made that most depictions involve creatures being hunted.

Cave art also provides insights into hunting techniques. Guthrie thinks that hunting parties averaged about five people, that the hunters were well clothed and that they may have used subterfuge (such as donning deer antlers) to get near prey. Spear wounds tend to cluster around the thoracic region, and often there is no spear shaft visible, suggesting that a spear with a socketed head was used. Moreover, the few images of the hunters themselves usually show each one carrying a single spear, for which they may well have carried multiple heads. Images also show single, injured animals, rather than herds. There is abundant evidence that ice-age Europeans used spear-throwers, some of which propelled fletched darts (darts with feathers at the rear of the shaft).[3] Fletched darts have the power to kill even when propelled from a great distance and are a highly sophisticated technology.

The fact that Guthrie dedicates his book to his boyhood mentors and friends might seem surprising—until you read that he thinks that most ice-age art was executed by feckless, idle youths. Analyses of handprints and finger smears left by the artists, mostly at sites away from the great galleries, suggest that the majority were made by young people, caught literally 'red handed' in the act of painting on cave walls. On occasion, the artists carried infants with them—the handprint of a very small child, along with the impression of its sleeve, is preserved at Gargas in France. Of a sample of 210 handprints, Guthrie determines that 169 were left by adolescent males and 39 by adolescent females or males aged between eleven and seventeen. A thorough study of the much more limited number of footprints gives a similar result. One engraving on a stone preserved at La Marche depicts a gang of four teenage boys, facial fuzz and all—perhaps self-portraits.

Many discoveries of ice-age art have been made by young people, including the galleries in Altamira, which were found by an eight-year-old girl, and the Lascaux galleries, which were discovered by eighteen-year-old Marcel Ravidat. It's the young who have the greatest spirit of adventure—and the size and suppleness to explore dark crevasses and caves, so it may be no coincidence that the artists and the discoverers fall into the same age group. Except at Chauvet, Lascaux and Altamira, most works are casual and improvised, replete with mistakes and clumsy executions.

Many of the less-sophisticated works are sexual in nature. Among the most common images are stylised vulvas, whole flocks of which appear on some cave walls. Less common, but still frequent, are erect penises, more complete female nudes, copulations, and even scenes of bestiality. We can imagine the circumstances. It's winter—frozen outside—and in the confines of a cave, mum and dad are being driven crazy by the high spirits of a group of bored adolescents. After a few harsh words, a youth grabs a torch and, taking a favourite baby brother with him for company, disappears with his gang into a crevice at the back of the cave, wherein lies a magic world, in which for a short time they can let off steam, and draw.

Some ice-age art remains enigmatic, including objects that resemble life-sized, erect penises carved from ivory, antler bone and stone. Were

they not so old, they might be identified as dildoes. A final feature of
Palaeolithic art that deserves comment is the large number of images
of full-figured women. Less than 10 per cent of all depictions of females
are of lean female forms, the remainder being described by Guthrie
as 'plump to corpulent'.[4] None, incidentally, have pubic hair. Guthrie
argues that it is likely that ice-age European women depilated their
pubic area (a practice common among tribal and western people today).
He thinks that these images (along with the countless disembodied
vulvas, some of which have been described as reindeer feet by prudish
researchers) are the work of males who were depicting what interested them sexually. Supporting his argument, Guthrie notes that there
are no females depicted wearing anything but rudimentary clothing
(though hair-styles are shown), and where men are drawn (there are
few images) they are clothed. Moreover, there are no depictions of
infants, pre-pubescent girls or post-reproductive-age women.

Guthrie argues that ice-age art provides an accurate picture of the
habits and appearance of those common large mammals that the artists
were dependent on for their existence. The hunters brought meat home
(often a cave) where it was shared, allowing women who were not
lactating to wax fat on the bounty. The creation of Palaeolithic art
was a largely male activity and much of it originated in ways similar
to modern graffiti. It is a view of European ice-age art that is earthy
and familiarly human, making the minds and culture of our distant
ancestors readily accessible.

Despite the great consistency in ice-age art over the millennia,
the relationship between animals and human hunters was changing.
Using the ghostly outline of vanished cultures that is the archaeological
record, we can make some guesses as to how. Spear points have left
a continuous record of rapid technological and cultural development
in Europe. Indeed, cultures have been characterised by their spear
points. The culture of the human–Neanderthal pioneers, known as

the Aurignacian (named after an archaeological site in France), was brief, lasting only a few thousand years. While the Aurignacians were capable hunters of the largest mammals, they were not equipped with the specialist flint points characteristic of later European cultures. Instead, they made finely crafted bone points to haft onto their spears, reserving the use of flint mostly for blades and scrapers.

Bone points work differently from those made of flint. Bone can penetrate hide and muscle, and a well-placed strike can cripple an animal if not kill it. But a misplaced hit will allow the prey to escape. Unless it can be tracked and retrieved, it is likely to die some time later of sepsis, out of range of the hunter. About 33,000 years ago, an important innovation in spear-point manufacture occurred. The Gravettian culture (again named for a site in France) flourished across Europe for nearly 10,000 years—until about 22,000 years ago—and its signature innovation was the development of a small, pointed flint blade with a straight, blunt back. It was a specialist tool used to hunt large mammals, including horses, wisent and mammoths, and was capable of causing death through blood loss, which is quicker than death through sepsis. Substantial blood flow had the additional benefit that the wounded beast leaves an ample spoor.

But spear-point innovation did not stop there. In France, northern Spain and possibly Britain, the Gravettian culture was succeeded by the Solutrean culture (named after a fossil site in southeastern France). Among its many achievements are the magnificent art galleries of Lascaux and Altamira, and the development of the eyed needle, which must have revolutionised the making of clothes and thus enhanced the ability to hunt in extreme weather. But the culture is best known for its spear points, which are renowned for their outstanding beauty. Solutrean points were made from flint and other stone selected for its aesthetically pleasing colour or patterning. The points were exqui-sitely shaped by sophisticated napping—doubtless the work of master craftsmen—being finely worked on both sides to have long, sharp cutting edges.

Solutrean points resemble the canines of the sabre-toothed cats. Indeed, they may have killed in a similar way—by exsanguination. They closely resemble the famed American Clovis points, which are associated with large mammal extinctions across the North American continent. Clovis points were made for just 300 years or so, with production ceasing at about the time the American megafauna disappeared: once the mammoths were gone, people stopped making the points used to hunt them. The manufacture of Solutrean points lasted about 5000 years, but by 17,000 years ago, with the woolly mammoth and woolly rhino in retreat in western Europe, they ceased being made.

The difference in timespan between the manufacture of Clovis and Solutrean points is intriguing. North America's mammoths had no experience of being hunted by hominids until highly armed human hunters arrived, after which the great beasts were exterminated quickly. Europe's mammoths, in contrast, after millions of years of hunting by *Homo erectus*, Neanderthals, humans and hybrids, were wary of upright creatures bearing sticks.

Why then did Europe's mammoths finally succumb? One answer may lie in the superior speed of cultural evolution over physical evolution. It took millions of years for the canines of sabre-toothed cats to become as large as they eventually did. But it took human spear points just 20,000 years to go from the Aurignacian bone spearhead to the more-deadly Solutrean point. The Red Queen hypothesis defines evolution as a kind of arms race, in which species must constantly evolve and adapt merely to survive. If you can't evolve fast enough, you become extinct. Mammoths and sabre-toothed cats evolved at the same pace, so the evolutionary arms race was kept in equilibrium. But when modern humans began their great cultural acceleration, large, slow-reproducing prey species had no way to keep pace.

While this account provides a satisfying narrative, there is a problem with the idea that Solutrean points were the deathknell for Europe's mammoths. A survey of the points found in Spain shows that very few bear fracture marks typical of flint spear points that have been used in hunting. Isabel Schmidt of the University of Cologne believes that this is because Solutrean points were largely symbolic, and not used for hunting.[5] There are other instances of such phenomena. The great and majestic Hagen axes of Papua New Guinea are exquisitely wrought and of great value. But they are never used as tools, serving instead to signal the status of their owners. But if Solutrean points did not often end the lives of mammoths and other megafauna, something else did. By the time Solutrean points were being manufactured the creatures were disappearing and given the many past shifts in climate that they had survived, climatic factors alone cannot have been responsible.

The striking similarity between Clovis points and Solutrean points has given rise to an odd theory. Some researchers posit that Solutreans pre-empted the Vikings in crossing the north Atlantic and colonising North America. But no other line of evidence (including genetic studies) supports this. And the dates are wrong. Solutrean points were made between 22,000 and 17,000 years ago, and Clovis points for about 300 years 13,000 years ago. It seems more likely that humans in Europe and North America hit upon a similar solution to the same problem—how to kill great, dangerous, hairy beasts quickly and efficiently—even if in the European case the fine tools eventually took on a largely symbolic value.

A further enigma concerns the disappearance of Europe's original mammoths about 34,000 years ago—several thousand years before the North American mammoths make their appearance in Europe. It may be that we just don't have enough samples to tell the full story. But it is intriguing that the extinction occurs at the time when the first Gravettian points were being made—33,000 years ago. Did the

Gravettians drive Europe's last indigenous mammoths into extinction at the ends of their lethal flint points, only to have mammoths of American origin replace them a few thousand years later? And did the Solutreans do the same for the last of southwestern Europe's mammoths? Given the deficiencies in the fossil record and the lack of focused studies, we cannot know with any certainty. But the patterns are tantalising.

Solutrean points were not made Europe-wide, but were restricted to a region extending from southern England to Spain. About 17,000 years ago the Solutrean culture was replaced by the Magdalenian culture, named for a rock shelter in the Dordogne where their artefacts were first recognised. The Magdalenians hunted a wide variety of prey, including horses, aurochs and fish, and are known for their highly sophisticated bone artefacts, as well as for small flint tools known as microliths, which were mounted together on spears to form a long cutting edge. It is during Magdalenian times that the last of western Europe's mammoths vanished, and when dogs began to be buried with people. The rapidly evolving Magdalenian culture would, in its many local manifestations, continue until the advent of agriculture.

IV

HUMAN EUROPE

38,000 Years Ago to the Future

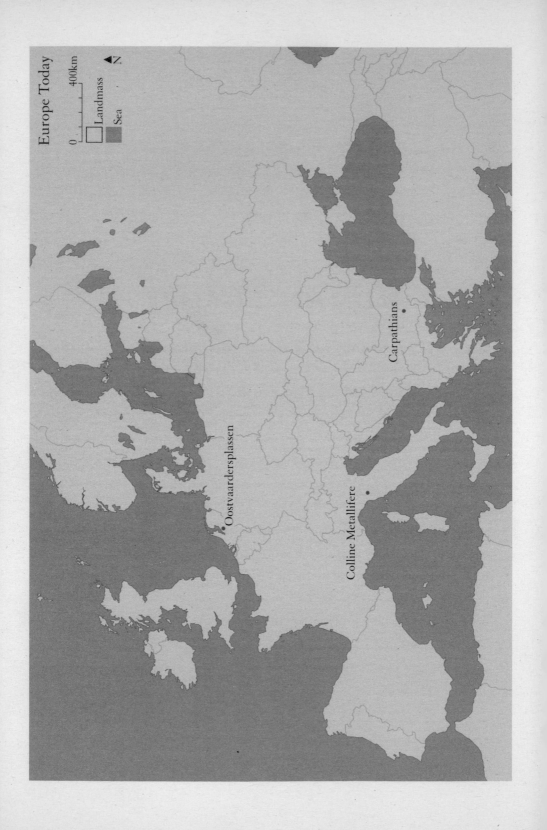

Europe Today

0 400km

Landmass
Sea

N

Oostvaardersplassen

Carpathians

Colline Metallifere

CHAPTER 32

The Balance Tips

Rhinos established themselves in Europe more than 50 million years ago, and elephants arrived 17.5 million years ago. They had endured everything from the Messinian crisis to the Anglian glaciation, but beginning about 50,000 years ago they started vanishing, and by about 10,000 years ago they were all extinct on the European mainland. For a century some scientists have waved airily at 'a change in the climate' as the cause. But things are not so simple. After decades of research, scientists have assembled an approximate chronology of extinction. On the European mainland, the straight-tusked elephant vanished sometime after 50,000 years ago, though some island populations survived until at least 10,000 years ago. While its extinction on the mainland appears to predate the arrival of humans, Signor-Lipps urge caution.[*] Europe's two woodland rhinos—Merck's and the narrow-nosed—also appear to have become extinct early. Again, we have very few accurately dated fossils, but if the depictions in Chauvet Cave represent a woodland rhino, then it must have survived until at least 37,000 years ago.

---

[*] An elephant depicted on the lion panel in Chauvet Cave is hairless and strangely thin. Although key features are ambiguous, it's just possible that it represents a straight-tusked elephant.

The Neanderthals became extinct about 39,000 years ago, after briefly overlapping with human–Neanderthal hybrids. The woolly rhino appears to have become extinct about 34,000 years ago. Next was the cave bear, whose well-documented extinction, at least in the European Alps, occurs about 28,000 years ago. The few dates we have for the cave hyena indicate that it was the next to go, about 20,000 years ago, and then, about 14,000 years ago, the cave lion vanished from Europe. About 10,000 years ago the last European woolly mammoths and giant elk become extinct, while the muskox died out (in Sweden) 9000 years ago.

EXTINCTION TIMES FOR THE EUROPEAN MEGAFAUNA

Animal	Extinction (years ago)
Mainland straight-tusked elephant	50,000
Neanderthals	39,000*
Merck's rhino	37,000
Narrow-nosed rhino	37,000
Woolly rhino	34,000
Cave bear	28,000
Cave hyena (survives in Africa)	20,000
Cave lion	14,000
Woolly mammoth	10,000
Giant elk	10,000
Muskox (survives in the New World)	9000

Some of these dates may be corrected as more research is done. But the pattern of extinction is not what one would expect if a changing climate was responsible. The most severe cold of the entire

---

* Dates are given without the uncertainty limits, which can span several thousand years. Thus, the most likely date for human arrival in Europe is given as 38,000, and the extinction of the Neanderthals as 39,000 years ago. In reality both events could have occurred between 35,000 and 40,000 years ago.

glacial cycle occurred 30,000–20,000 years ago, before the ice began collapsing about 16,000 years ago. The large warmth-loving species, including the straight-tusked elephant and the woodland rhinos, should become extinct at the time of maximal cold—but they go earlier. And cold-loving species should become extinct as the warming sets in; but the woolly rhino dies out earlier, while the muskox survives longer. A second curious feature is that only the largest mammals become extinct.* In today's rapidly warming climate, small creatures such as pikas and saiga antelope, as well as large ones, are declining.

With the extinction of the mammoth about 10,000 years ago, mainland Europe had lost all herbivores that weighed more than 1.5 tonnes, and all its strict carnivores weighing more than 50 kilograms. Megafauna are called 'keystone species' because when they become extinct the entire arch of the ecosystem can collapse. In the case of the woolly mammoth and the mammoth steppe, the collapse fits that expectation and has been documented. But elsewhere in Europe, evidence for large-scale ecosystem collapse is lacking. An idea of what should have occurred is provided by Africa: where elephants have been hunted out, the savannah transforms into woodland or even dense forest, forcing all the smaller savanna-living creatures to find habitat elsewhere. In Europe, thick forests briefly re-established following the final retreat of the ice, but within a few thousand years fire-using and axe-wielding humans were replacing elephants as disturbers of vegetation.

Large predators are equally important as keystone species; their removal can allow predators in the next size class down to proliferate. This phenomenon, known as meso-predator release, can have a huge impact on ecosystems. Barro Colorado Island in Panama became an

---

* The small creatures intimately associated with them, such as the dung beetles, also become extinct in Europe. But, generally, the smaller European vertebrate species that inhabited Europe during the glacial maxima find refuges somewhere in Europe.

island through the flooding of the Panama Canal. It was too small to support the region's top predators, jaguars, and the big cats died out. Afterwards, smaller predators became super-abundant and caused the extinction of several bird and mammal species that were important pollinators and seed dispersers. This, in turn, altered the composition of trees in the forest. But Europe's medium-sized carnivores remained relatively rare after the extinction of the great predators. Again, humans are implicated. We see in archaeological deposits that humans were becoming expert hunters and trappers of carnivores. In effect they were replacing lions and hyenas as suppressors of medium-sized carnivores.

Another consequence of the extinctions can be expected: as in ages past, new kinds of elephants and rhinos should migrate into Europe to replace the extinct types. After all, as the warming set in, the habitable land area of Europe increased enormously, and with greater rainfall and soils revitalised by glacial activity there was increased biological productivity. In previous ice ages these factors triggered mass migrations of mega-mammals into Europe. But at the end of the last ice age the mega-herbivores did not return. The only obvious explanation is that the density of skilled human hunters in Europe was preventing megafaunal recolonisation.

It seems that long before the advent of agriculture, humans had replaced the ecosystem function of all the great beasts of the ice age. By about 14,000 years ago, Europe was already a human-maintained ecosystem. Indeed, the number of humans in Europe had begun to increase greatly. A recent study estimates that 23,000 years ago the human population of Europe was about 130,000, and by 13,000 years ago that number had more than trebled to 410,000.[1]

Despite the human influence, about 10,000 years ago some surprising invaders arrived in Europe. Western Europe, at least, had been lionless for nearly 5000 years, following the extinction of the cave lion, when prides of a new type of lion stalked in from Africa or western Asia. These were *Panthera leo*, the only lion species surviving today.[*] By 10,000 years ago it had reached as far west as Portugal, having colonised France and Italy on the way, and it survived in Iberia for at least 5000 years.[2] But as human populations grew, it was pushed eastwards. By Herodotus's time, west of the Bosphorus lions were plentiful only on the plains of Macedonia and by the last century BCE they had vanished even from there. They survived in Georgia until about 1000 CE, and in eastern Turkey until the eighteenth century; which, incidentally, allows them to fit the exacting standards of the IUCN as candidates for reintroduction to Europe![3]

Another surprise invader was the striped hyena. Just as the lion had to some extent usurped the niche of the cave lion, so the striped hyena partially usurped the role of its much larger relative, the spotted hyena. Remains of this species are unknown from Pleistocene deposits in western Europe. It may have migrated out of Africa as recently as Neolithic times (10,200–4000 years ago) and briefly became common in western Europe, particularly in France and Germany, before declining to extinction everywhere except the Caucasus, where it maintains a precarious toehold.[4]

While animal migrations were few, the pace of human migration picked up. Genetic studies, including those of fossil DNA, show that about 14,000 years ago a group of humans began expanding westwards from what is now Greece and Turkey (though they may have originated further east). They mixed into the European gene pool, and probably displaced some of the original settlers. Consequently, the average amount of Neanderthal DNA in European populations dropped to about two per cent.[5]

---

[*] Although we think of lions as being African, until recently they were widespread in western Asia, a relic population of which survives in western India.

There is evidence that these new people differed in important ways from the original inhabitants, and some telling insights into their culture have been unearthed in Turkey, at Göbekli Tepe, where the world's oldest 'temple' has been discovered. The Göbekli Tepe temple was constructed some 11,500 years ago, a few thousand years after the initial spread of the new migrants into western Europe, but prior to the advent of agriculture. It is likely that the ancestors of the builders of Göbekli Tepe shared a common culture with the migrants who entered Europe at least 14,000 years ago.

The temple at Göbekli Tepe appears to be very different from anything that preceded it, but this may be owing to a bias in preservation. Archaeologists refer to the classical Greek temple form as 'petrified carpentry', because stone temples arose from earlier, wooden types, and elements of carpentry techniques are retained in the stone forms. There is doubtless a long tradition, in many cultures, of construction in wood prior to the adoption of stone, and it is to be expected that the builders of Göbekli Tepe were no exception. We can thus anticipate that some of the adaptations seen in the culture that gave rise to Göbekli Tepe had been present in their ancestors at the time they expanded into western Europe.

The temple at Göbekli Tepe is an immense circular form, constructed of stone pillars up to six metres tall and 20 tonnes in weight, which were decorated with relief carvings of animals and stylised humans. This style of carving, in which the image stands out from a flat background, is markedly more difficult than simply incising a design into the stone. It is, however, often used in woodwork, where it is easier to execute, and it seems possible that its use at Göbekli Tepe constitutes an example of 'petrified carpentry'. The function of the Göbekli Tepe complex is still debated, but the discovery of human cut-marked skull fragments, as well as the bones of birds of prey, suggests that it may have been a place where the bodies of the deceased were exposed, to be eaten by vultures. A remarkably similar practice survives today among India's Parsees, who leave their dead in 'towers of silence'.

The construction of Göbekli Tepe would have demanded a considerable workforce. Among the major tasks were the preparation of the ground, the quarrying and transport of the columns about 800 metres, and then their carving and erection.[6] Just how the workers were fed is not known, but it must have required a food abundance. The archaeologists who excavated the site think that the gazelle and aurochs' bones found in the backfill used to bury the structure are evidence of great meat feasts. But it's hard to believe that sufficient meat could have been procured from hunted game to sustain such a workforce. Instead, it seems likely that they also ate some type of storable plant food in the form of seedcake or nuts.

Interpretations are necessarily speculative. But it seems possible that the builders of Göbekli Tepe had already embarked on an important phase of development that was a precursor to domestication, and which requires the manipulation of wild resources. This may have involved the selective planting in accessible locations, and the nurture of seedlings of fruit or nut trees that bore especially well. Such trees can take decades to produce large crops, so it's fair to ask why anyone would plant them if they're unlikely to live long enough to receive the benefit. But a similar practice continues in Papua New Guinea, where people plant trees that, when mature, are known to attract game animals. I have asked New Guineans why people do this, and they tell me it's so that their grandchildren will have food.

Any trees planted by the builders of Göbekli Tepe would not have had a high chance of survival if large numbers of animals browsed nearby, so herbivores must have been scarce. The structure occupies a hilltop, which is a great vantage point from which to spot migrating game. It seems possible that the people lived around Göbekli Tepe for a significant part of each year, and that they exerted severe hunting pressure on local animal populations in the vicinity. The suppression of herbivores could have opened another resource—grass seeds. Where they're grazed, grasses reproduce asexually using underground rhizomes, but when grazing is reduced grasses reproduce sexually from seed heads. Grass seeds are a key human resource, for they are nutrient dense and can be stored, or ground into flour and cooked into storable cakes, which can be fed to workers.

This hypothetical economy of the Göbekli Tepe people, however, leaves us with a problem, for in places like Turkey grasslands can quickly turn into forests if grazers are absent. To prevent this, and keep the grasses productive, the builders of Göbekli Tepe could have used fire like Australia's Aboriginal peoples do. The judicious use of fire could have protected valuable nut trees, promoted the growth and seeding of grass and, if practised at a distance from their camps, even attracted herbivores to the sweet young pick. One enormous benefit of using fire in this way is that it allows advance planning. Climatic conditions permitting, it can be anticipated that game animals will come to forage on the regrowing grass a certain number of weeks after burning, and that grass seed will become available after a longer interval.

This all adds up to a kind of 'proto-domestication' stage in the management of wild resources, involving significant ecosystem manipulation, but not planting or intensive selection of plants for seed size. This would make the people who pushed into western Europe about 14,000 years ago quite different from the low-population-density living, big game hunters they replaced. Skilled landscape manipulators who were harvesting from low down the food chain, the newly arrived groups could build up dense populations, as well as provisioning individuals to build temples—or wage war.

CHAPTER 33

The Domesticators

The human population of Europe must have grown between 13,000 years ago (when there were an estimated 410,000 Europeans) and 9000 years ago, as the climate had warmed and stabilised. The retreat of the ice revealed newly created or rejuvenated soils across northern Europe, which was then rapidly colonised. In effect, a great new fertile land, enjoying an increasingly warm climate, was inherited by the most robust of pioneering plants and animals. Among the first movers were the surviving tundra species—the lichen and the reindeer that fed on them, the dwarf willow, the mountain hare, arctic fox and lemming. Then came the mixed deciduous forest, which expanded rapidly, and by about 8000 years ago had reached its current extent.[1] Among the adaptable colonisers that thrived in it were the red squirrel, hedgehog, red fox and badger—all of which were destined to become familiar creatures in modern Europe, and some of which would travel with the Europeans during the age of empire to faraway lands, where they would entrench themselves as pests.

Of course, larger creatures were present as well, including bear, wolf and red deer. But it was the humans who were set to prosper most in the newly habitable lands. The agriculturalists arose about 11,000 years ago, somewhere in a swathe of country extending from present-day eastern Turkey to Iran. Goats, sheep, pigs and cattle became domesticated more or less simultaneously, and the process of domestication must have been deliberate, for someone—possibly children—must have tended the flocks, taking them daily to pasture and returning to the encampment at night.[2] How could that have begun? Dmitry Belyayev's fox experiment tells us that the traits seen in almost all domestic species develop through selection for tameness. We can imagine that, over the millennia, many baby herbivores were brought into campsites, and those that were calmer in the presence of humans would have 'self-selected' to stay in the camp after reaching sexual maturity. But why was it only from 11,500 years ago that this led to domestication?

It's widely believed that agriculture and animal husbandry developed because of climatic stabilisation about 11,000 years ago. The argument has been put best by Brian Fagan in his book, *The Long Summer: How Climate Changed Civilisation*.[3] Fagan argues that ice-age climates were hostile, but that an exceptional period of climatic stability that followed allowed agriculture to be worthwhile. Climate clearly has an impact in agriculture, but I think it is a mistake to think of it as the only factor, or even the decisive factor. During the ice ages, climatic change was far slower than it is today; shifts in climate would have been imperceptible to the people living back then.* Because large areas at low latitude were suitable for agriculture even during the ice age, we must look more widely for an answer. Perhaps, as Fagan says, ice-age weather was wilder and more disruptive than that which followed. But this is yet to be demonstrated to a convincing degree.

When searching for the origins of domestication, it's helpful to think of it as a form of delayed gratification, for grain must be sowed for later consumption, rather than eaten now, and a flock must be built

---

*Climate change is progressing today at least 30 times faster than at any time during the past 2.6 million years.

up before it can be culled. For this to be worthwhile, there must be a reasonable prospect of return. A hostile climate can certainly limit the size of the return, and wild weather can destroy it. But the return is also dependent upon factors that humans can control. For example, herbivores can destroy a crop, or predators ravage a herd, while hostile neighbours can destroy or steal both.

The hunting and fire-management skills of the Göbekli Tepe people would have increased the prospect of a return on the investment involved in agriculture. But to make the bargain truly worthwhile, they would need to prevent neighbours from raiding their flocks and crops. We know little of their political organisation, but it's not unreasonable to think that it may have taken a thousand years or so after the construction of Göbekli Tepe for them to acquire such refinements.

Analyses of animal bones from the earliest village sites reveal that people were selectively culling young males from their herds. This process of 'unnatural' selection allowed whatever it was that caused people to spare the odd male from the butchers' knife, to survive into the next generation. Within a few thousand years those selected traits would give rise to the various domestic types that we see in the archaeological record.

Goats, sheep and pigs were all important to the earliest European herders, but how much more significant was the cow to become! Hindus retain an appropriate reverence for the creature. Not so the Europeans—a people mythologically conceived from the coupling of a bull and a goddess: Europa, the wide-eyed one, the cow-faced one, was abducted by a white bull—the god Zeus in disguise—and she bore him three sons. Echoes of the centrality of cattle in European culture are to be seen in Neolithic engravings of oxen pulling carts that are carrying the sun. Even today, in Sienna in Tuscany, during the Palio parade, oxen pull carts through the city streets—in what is perhaps a survivor of an Etruscan cultural practice.

As its mythology suggests, Europe owes more to the union of cow and human than any land ever did. Europeans are among the few who can wax fat on the milk of the cow, having the highest lactose tolerance on Earth—the Irish being champion of champions in this regard. The ability of Europeans to thrive on milk has, of course, been built on the misfortune of countless of their ancestors—who could not tolerate lactose as adults. They perished, so that we lucky few could build a civilisation on the strength of the cow.

At domestication's dawn, I have no doubt that the cow was considered a member of the family, a creature protected and cosseted, which in return gave nourishment. Anyone who has hand-milked a cow will know that, for three weeks after the birth of her calf, her milk is especially delicious and rich. But woe betide the person who tries to milk her before her calf has had its fill. She will resist with all her wiles—as is her right in the bargain of domestication we have struck with her—before giving in to the human who wishes to taste the delicious drop. Yet, after her calf has drunk its fill, she will contentedly yield her udder, even taking a slurp herself if you aim the stream at her mouth.

Today, the cow is no longer regarded as a family member, but a unit of production. She is often miserable, perpetually confined to her stall in a mechanised dairy factory. The udders of her aurochs ancestor were so small that even when lactating they were barely visible. But over thousands of years of unnatural selection, the udders of the dairy cow have grown so prodigious that today their weight often cripples her legs, and these bovines are prone to the potentially deadly disease mastitis. It is not, it seems to me, worth reneging on the deal we first made those thousands of years ago, for the sake of a glass of cheap milk.

The earliest domesticators were sailors who settled some of the larger Mediterranean islands. In mainland archaeological sites the bones of early domesticates are inevitably mixed with those of wild animals. But on islands there were no wild sheep, goats or cattle, making interpretation easier. Cyprus, which is 60 kilometres from the Turkish mainland, was discovered by early domesticators about 10,500 years ago. They brought with them sheep, goats, cattle and pigs that did not differ in body form from their wild ancestors, but whose remains indicate that the young males were being culled.

One of the most modified domesticates is the sheep, the various breeds of which bear little resemblance to its wild ancestor, the mouflon. Remarkably, the descendants of some of the very earliest domestic sheep roam wild today on Corsica, Sardinia, Rhodes and Cyprus (where the local population has become so distinctive that it is recognised as a subspecies). Their ancestors must have escaped soon after the first domesticators and their herds arrived between 7000 and 10,000 years ago.

The early domesticators also brought the fallow deer and fox to Cyprus. Perhaps the young of these creatures had been adopted by children and had escaped when they reached the island. Some researchers, however, think that they were deliberately released to stock the island with game. The settlers also brought crops—in the form of einkorn wheat, emmer wheat and lentil—and began farming. By 7300 years ago, the domesticators and their flocks had spread from their point of origin in the Levant to coasts as far as western Iberia.[4] As sheep and goats spread westwards, they did not encounter similar native species with which they could hybridise. But domestic pigs would have encountered herds of wild boar, and domestic cattle would have encountered aurochs. Genetic studies reveal that European wild boars bred with domesticated sows, their genes entering the domestic herd. This genetic influence would eventually spread back east, far beyond the geographic range of the European wild boar.[5]

Analysis of the genome of an aurochs that lived in Britain 6750 years ago revealed that her genes survive in some ancient British and Irish breeds. The ancient herders might have captured aurochs to add to their herds as they became depleted. The study also revealed that the genes regulating the brain and nervous system, growth, metabolism and the immune system have altered most in the modern breeds relative to their wild ancestors.[6] Domestication has also altered humans. The idea that we have undergone auto-domestication can be attributed to Dmitry Belyayev. One result is that today many human cultures value husbandry more than hunting: over 10,000 years, evolution has favoured those with the ability to nurture crops and herds. And even though we like to think of ourselves as the lords and masters of the farm, we have experienced very strong natural selection since farming began, resulting from a changed diet, changed exposure to disease, and a more sedentary lifestyle.

CHAPTER 34

From the Horse to
Roman Failure

Research into the genetics of late-surviving European hunter-gatherers and the newly arrived farmers suggests that the farmers almost entirely replaced the earlier occupants in many areas.[1] Since the advent of writing, Europe's human history has been a sorry tale of war and extermination, so the replacement of one people by another 8000 years ago is not surprising. An analysis of skeletons from cemeteries shows that for about 700 years after agriculture became established in any area, the population increased rapidly. This was followed by a stable period lasting about 1000 years, after which populations began to collapse, and within a few centuries human numbers became much reduced.[2] The expansion of the agriculturalists would not be the last great human migration into Europe. About 5000 years ago, horse-riding herders from the Russian steppes arrived in Europe, again displacing some peoples. As a result of this long history of invasions, going back all the way to Neanderthal times, every European living today is of widely mixed heritage, as our variable eye, skin and hair colour and form suggest.

Ever since humans first arrived in Europe, migration had been westward, and prior to the eighteenth century the great innovations were mostly eastern, often only penetrating Europe after considerable delays. A hundred years ago Europeans were all but unaware of this.

The idea that Europe was an appendix of Asia as far as human cultures were concerned would have been derided as ridiculous or considered insulting. Among those who delivered the news that Europe was indisputably not the cradle of civilisation was Vere Gordon Childe, the first and arguably the greatest ever synthesiser in archaeology. Famous for his view that European civilisation was a 'peculiar and individual manifestation of the human spirit', rather than the apogee of human achievement, he was also one of the greatest eccentrics to wield an archaeologist's trowel.[3]

Childe was—like that other original thinker Baron Nopcsa—the quintessential outsider. Born in Sydney, Australia, in 1892, and son of an Anglican reverend, Childe was a perpetual valetudinarian, so sickly that he was home-schooled, and of such an 'ugly appearance' that he became the butt of cruel jokes.[4] Awkward, uncouth and without social graces, Childe seems never to have had a sexual relationship.[5] His one love in life, outside his work, was speed. He owned several fast and expensive cars, and after he moved to England he became infamous for his reckless driving—including a high-speed dash down Piccadilly in the early hours of the morning that attracted the attention of the police. After winning a scholarship to Oxford he became frustrated that his Marxist views prevented him gaining an academic position. So he returned to Australia to work for the New South Wales Labor Party, and wrote a book called *How Labour Governs*—a highly insightful if disillusioned study of worker representation in politics.

After returning to London in 1922, Childe was unemployed for several years; but this was the most productive time of his life. He researched in the libraries of the British Museum and the Royal Anthropological Institute, and in 1925 published *The Dawn of European Civilisation*. Along with *The Aryans: A Study of Indo-European Origins*, which came out a year later, the book established conclusively the importance of the East as the fountainhead of 'European' civilisation.

As a Marxist historian, Childe saw prehistory in terms of revolutions and changing economies. Among his great excavations was Skara Brae, a famous Neolithic complex in the Orkney Islands, and among his profound insights was the identification of the 'Danubian Corridor' which was used by many species, including Europe's human–Neanderthal hybrids, to migrate westwards. A fervent supporter of the Soviet Union, in 1945 he wrote to his friend Robert Stevenson, the Keeper of National Antiquities of Scotland, that 'the brave Red Army will liberate Scotland next year, the Stalin tanks will come crunching over the frozen North Sea'.[6] The brutal Soviet suppression of the Hungarian revolution in 1956 disillusioned him, and towards the end of that year he retired prematurely from his position as director of the Institute of Archaeology and returned to Australia. In a letter dated 20 October 1957, marked 'not to be opened until January 1968', he told of his last days:

> I have always considered that a sane society would...offer...
> euthanasia as a crowning honour...For myself I don't believe
> I can make further useful contributions...An accident may easily
> and naturally befall me on a mountain cliff. I have revisited my
> native land and have found I like Australian society much less
> than European without believing I can do anything to better it:
> for I have lost faith in my ideals.[7]

On 19 October 1957 the great archaeologist flung himself over the 1000-metre-high precipice known as Govetts Leap in the Blue Mountains, near where he had grown up. We can only hope that he revelled in the acceleration of his last moments.[*]

New species continued to be added to the human retinue. The cat seems to have domesticated itself in the near east by 9000 years ago.

---

[*] Inexplicably, the jump was the day before the date on Childe's letter.

And around 5500 years ago the most important species to join the household—the horse—was domesticated somewhere in the steppes of western Eurasia. It arose from *Equus ferus*, a genetically well-mixed species (having little geographic variation) that existed across a vast range, from Alaska to the Pyrenees. Unlike the aurochs, whose regional ancestors can be traced genetically, the history of the horse is 'a genetic paradox', though it's clear that Przewalski's horse is not an ancestor of the domestic horse, but a separate lineage going back 160,000 years.[8]

There is so little variation in the Y-chromosome of domestic horses that the original herd must have had very few stallions. In contrast, mitochondrial DNA, which is only passed on through the female line, is spectacularly diverse. This could be because there were a very large number of mares in the original stock, or perhaps because additional mares from wild herds were added as domestic horses spread across Eurasia—an idea supported by the latest data. It seems that many were taken in during the Iron Age, between around 3000 and 2000 years ago.[9] In genetic terms, no living horse breed is a surviving representative of *Equus ferus*.

Very few species have been domesticated since the horse. The bee was domesticated in Egypt by 4500 years ago, and the dromedary about 3000 years ago on the Arabian Peninsula: it was nearing extinction at the time, being restricted to mangrove areas in Arabia's southeast.[10] About 3000 years ago the Bactrian camel was domesticated in central Asia, and the reindeer may have been domesticated in both Siberia and Scandinavia. The only more recent examples are the rabbit and the carp, which were domesticated by monks in the middle ages.

You might have noticed an omission in this tale of domestication— the Romans. Few people have ever had access to the variety of wild animals that the Romans did, or kept them for such an astonishing variety of purposes. From lions to elephants and bears, destined for combat in the arena, to the lions that Mark Anthony reputedly used to draw his chariot, wild animals were captured and trained en masse. If, incidentally, Anthony's lions were anything but legend, the feat of harnessing these big cats was one of the greatest triumphs ever of man over beast.

The Romans found dormice an irresistible delicacy, and to satisfy their appetites they would capture wild individuals and keep them in terra cotta containers known as *gliraria*, while they fattened. Dormice are only very distantly related to the rats and mice that infest our houses and crops; they are surviving members of Europe's most ancient mammal lineage, whose history stretches back more than 50 million years. Yet for all their expertise in other areas, the Romans never domesticated dormice: they never got them to reproduce in captivity— a key threshold for domestication.

The Romans were also famous for keeping fish, including red mullet, which were captured as juveniles and grown to enormous size in ponds. A large red mullet could cost as much as a slave. And the Romans were the first to cultivate oysters, the praetor Caius Sergius Orata, who grew them in the Lucrine Lake—a coastal lagoon in the region of Baiae (modern Baia)—in the first century BCE, was the first recorded oyster farmer.[11] But Orata's oysters were, like the dormice and fish, collected in the wild, as spat. So oyster farming, like dormouse-fattening, is not a form of domestication, but rather captive rearing.

The failure of the Romans to add to the domestic stock is truly inexplicable. For about 500 years they ruled a peri-Mediterranean empire that was about the same size as the Incan empire in South America—though it lasted five times longer. Situated in a biologically diverse part of the planet, which they scoured in search of wild animals, and having the advantage of Virgil's *Georgics* (an instructional poem dealing with agricultural techniques) and all their expertise in training, captive-rearing and even selectively breeding already domesticated creatures, they failed to domesticate a single species. Yet barbarians, whose cultures they knew and who lived just before them, as well as the Europeans of the middle ages who succeeded them, both added species to Europe's domestic stocks.

CHAPTER 35

Emptying the Islands

Islands are central to the story of Europe, and even today its numerous islands are diverse and ecologically important. Yet so very much has been lost: the fate of Europe's unique island faunas over the past 10,000 years is an extreme example of how natural heritage has been diminished by relentless human expansion. The story begins on Cyprus, the first major Mediterranean island to be colonised by humans. Whoever saw the island in its virginal state must have experienced a paradise.

Indications of what they encountered were unearthed by the founding mother of Mediterranean islands palaeontology, Dorothea Bate. Born in 1878, Bate received little formal education (she once quipped that her education had only briefly been interrupted by school). She became a 'piece worker' at the British Museum. The lowliest of employees, piece workers were paid only for each bird or mammal they stuffed, or each fossil they prepared. Bate persisted in this precarious occupation for more than 50 years, all the while teaching herself how to find fossils and research and write scientific papers.

She was fortunate to meet the Swiss palaeontologist Charles Immanuel Forsyth Major, who encouraged her to visit the islands of the Mediterranean in her search of fossils. Her first venture was to Cyprus, where she had been drawn by ancient accounts of bones in caves said to belong to the seven martyrs of island tradition, or the seven sleepers, who entered a cave and slept for a year. In 1901 she set out for the island on a self-funded expedition and stayed for eighteen months, locating several of the caves mentioned in older texts, including the 'Cave of the Forty Saints' at Cape Pyla, which contained rich deposits of fossil bones.

The fossils Bate excavated are now held in the collections of the Natural History Museum, London, where the most interesting remains long lay unstudied. But in 1972 Dutch palaeontologists Bert Boekschoten and Paul Sondaar announced that the bones came from an unusual, tiny hippo, which they named *Phanourios minor*—'small manifested saint'; the cave had been visited for centuries by villagers seeking the fossilised bones of their 'saint', who they believed could cure various maladies.[1] The saintly hippo was, at less than a metre high and weighing only 200 kilograms, an island dwarf, which had presumably descended from full-sized amphibious ancestors that had swum out of the Nile to Cyprus. The hippo was widely distributed across the island, and it looks to have become entirely terrestrial in habits. Lacking predators, it may have been slow-growing and fatally naïve about carnivores.

The hippo shared Cyprus with a miniature straight-tusked elephant. The presence of small elephants on various Mediterranean islands may have influenced classical mythology. In 1914 the fascist Viennese palaeontologist Othenio Abel (he who disparaged Nopcsa's island theory of dinosaur evolution) proposed that the fossilised skulls of dwarf elephants might have given rise to the story of the cyclopses— the one-eyed giants of Greek mythology, which appear in different guises in several stories. In the *Odyssey,* the cave-dwelling cyclops Polyphemus, who is said to live in a 'distant country', which is often taken to be an island, captures Odysseus and his crew. Kept for eating, they escape by blinding the giant. The skulls of dwarf elephants, Abel observed, are about twice the size of a human skull, so might have been thought of as belonging to giants. Moreover, they have a central nasal opening that could be mistaken for an eye socket. The discovery of such a skull in a cave, he thought, might have triggered the tale of the cave-dwelling cyclops.

Malta and Sicily were once joined, so they share a common biological heritage. But by the time humans arrived they had been separated for hundreds of thousands of years. Sicily lies close to the mainland, the Strait of Messina hardly posing a barrier for many large mammals, including aurochs, wisent, red deer, ass, horse and straight-tusked elephant, all of which swam to the island. Sicily's straight-tusked elephants, incidentally, survived there until about 32,000 years ago, while the rest of the large mammals perished following the invasion of the island by the domesticators and their livestock.

Malta has a rich and varied faunal history, including dwarf elephants and hippos, and a giant, flightless swan that stood taller than the island's pachyderms. But by the time humans discovered Malta, it had a more limited fauna, perhaps because various species had been lost due to the small size of the island during times of high sea levels. The survivors included several kinds of dormice and a deer, both of which were unique to Malta, and neither of which survived the human impact.

Sardinia and Corsica are large islands that were connected at periods of low sea level. By the time humans arrived about 11,000 years ago, the islands' fauna included a dwarf mammoth, a deer, a giant otter with broad crushing teeth that probably fed on shellfish, three other otter species, a pika, some rodents, shrews and a mole. It was also inhabited by a small dog-like canid known as *Cynotherium*, which may have fed exclusively on the island's pikas.[2]

About a million years ago, Sardinia and Corsica were settled by a *Homo erectus*-like species that left abundant stone tools. But it became extinct and thereafter the islands lacked upright apes until they were rediscovered by the domesticators. The mammoth and other large mammals seem to have become extinct shortly thereafter, but the deer survived until about 7000 years ago, and the pika right into the eighteenth century (where it persisted on offshore islets). Sadly, today the entire endemic fauna is extinct.

The Balearic Islands of Minorca and Majorca were not discovered by humans until between 4350 and 4150 years ago—around the time of Egypt's Old Kingdom—and so managed to hold onto their unique fauna for six millennia longer than the more easterly Mediterranean islands: three very unusual creatures inhabited these islands—the giant shrew *Asoriculus*, the giant dormouse *Hypnomys* and *Myotragus*—an enigmatic member of the goat family. The shrew is little known, but the dormouse weighed up to 300 grams and was probably partly terrestrial (rather than arboreal) and omnivorous.[3] *Myotragus* (meaning 'mouse-goat') weighed between 50 and 70 kilograms and browsed bushes. The remains of all three of these strange creatures were, after many unsuccessful searches and dead ends, discovered and named by Dorothea Bate, who also published a brief description of the mouse-goat in 1909.

Bate's work was not without its hazards: she caught malaria on Cyprus and nearly starved on Crete. On Majorca she was sexually harassed by the British Vice Consul, writing of the experience: 'I do hate old men who try to make love to one and ought not to in their official positions'.[4] Bate was a strong personality, and I suspect that her wry sense of humour was responsible for her idiosyncratic wording. Bate, incidentally, did not limit herself to fossils; the modern spiny mouse of Cyprus is among her discoveries. And, when in her 70s, she discovered the remains of a giant tortoise in—of all places—Bethlehem!

The ancestors of *Myotragus* seem to have walked to the Balearics nearly six million years ago, during the Messinian crisis, when the Mediterranean Sea dried up. Isolated for millions of years, they developed some highly unusual characteristics. Their eyes faced forward like those of monkeys and cats, rather than the usual sideways-oriented eyes of herbivores. And, like rodents, mouse-goats had a single stout incisor at the front of each lower jaw (hence the name mouse-goat). Their bones appear to have grown differently from those of any other mammal. Like the bones of reptiles, they have lines within them indicating long periods during which no growth occurred and the animal seems to have ceased much metabolic activity. This has prompted scientists to think that mouse-goats entered a kind of hibernation, or aestivation, perhaps in response to a lack of food or water. The young were very large at birth, and independent at an early age. Sometime after 4800 years ago, about the time the first humans arrived on the islands, the last mouse-goats died. It was once thought that the first humans in the Balearics had domesticated the mouse-goat, as what appeared to be pens filled with dung were found in some caves. But further study revealed them to be natural features.[5]

And so, beginning with Cyprus' dwarf elephants and hippos, did the unique species of the European islands vanish, until even the rat-sized Sardinian pika, which survived into Roman times and probably beyond, was driven into extinction by hunting or competition with species humans carried to the islands.* Today, among all of the Mediterranean's unique island mammals, there is a single survivor— the Cypriot mouse (*Mus cypriacus*)—a species so obscure and small that it was not even recognised as being distinct from the house mouse until 2006.** Was there ever a sorrier story of human ignorance and

---

* The Sardinian pika (*Prolagus sardus*) weighed around half a kilogram. A burrower, it abounded at the time of human settlement. It may have survived on the island of Tavolara, off the northeast coast of Sardinia, until 1780, when that island was finally colonised.

** The spiny mouse of Cyprus may also survive, though it appears to be on the brink of extinction, and two endemic shrews also survive on the Canary Islands in the Atlantic. Other island populations of shrews and mice, which are occasionally claimed to be species unique to islands, may in fact be descended from recent immigrants.

overexploitation than this? That every island from the coast of Turkey to the Pillars of Hercules has been emptied of its natural treasures, one by one, until just a single mouse was left.

CHAPTER 36

The Calm and the Storm

After the last European muskox died in what is now Sweden about 9000 years ago, the European mainland did not lose another species until the seventeenth century. In light of the changes that were occurring in human societies, this extinction gap is utterly extraordinary, for Europe's human population increased 100-fold, and the Europeans transformed from hunter-gatherers into farmers, bronze and iron tools were invented, and social organisation went from the level of the clan, to that of the Roman Empire.

Extinction is merely the final act in what is usually a drawn-out process. During the extinction gap Europe's larger mammals remained under relentless and increasing pressure from hunters and competition with domestic livestock. With each passing millennia their distributions became more restricted, their last refuges in areas unfavourable for human occupation, and perhaps in borderlands between tribes. Once the extinction wave broke, in the mid-seventeenth century, it quickly gathered strength, sweeping away the last survivors of group after group.

As with earlier extinction waves, this one disproportionately affected the largest species, but so severe was it that even the Eurasian beaver, which once thrived in waterways and lakes from Britain to China, was all but exterminated; there were just 1200 surviving globally by the early twentieth century. Historical records make clear that the cause was an increasingly dense, and deadly, human population.

By 200 CE the population of the Roman Empire (which then included much of Europe, along with parts of North Africa) was about 50 million people—that is 100 times greater than the population of the Europe of 11,000 years earlier. Importantly, in Roman times, between 85 and 90 per cent of people lived outside cities, surviving on what they and their communities could grow or catch.[1] By 1700, Europe's population had approximately doubled from what it was 1500 years earlier, to about 100 million—and the percentage of people living outside cities had hardly changed.

During the next two centuries, between 1700 and 1900, Europe's population quadrupled, to 400 million. Yet, except in industrialised Britain (where the proportion of people living outside cities had dropped to around 75 per cent), 90 per cent of Europeans still lived outside cities. By the first half of the twentieth century, almost every available scrap of land, with the partial exception of the royal hunting reserves, was being squeezed for every ounce of productivity it could yield. In Mediterranean Europe, sheep and goats by hundreds of millions roamed the mountains, consuming all kinds of vegetation, and wherever possible hills and mountains were terraced for cultivation.

One important factor prevented this great human expansion from destroying even more species than it did. It arose from a peculiarly European attitude towards hunting. In Roman times, hunting was mainly carried out by servants and slaves. But by the Middle Ages it had attained symbolic meaning and was part of a complex social system. The *caccia medieval* restricted the hunting of certain creatures to particular social groups. It quickly became widespread across most of Europe and remained substantially unchanged until the French Revolution. The *caccia medieval* reserved to the landlords and their families the hunting of red deer, wild boar, wolf and bear—the noble game. The smaller game, such as hares and pheasants, was usually left to the servants and common farmers.

It was from the *caccia medieval* that the large game reserves of most of Europe sprang, and in some places they remained until the end of World War II. One of their most ardent proponents was Spain's Alfonso XI (1311–1350). Renowned as a skilled hunter, he wrote the popular *Book of Hunting*, in which he tells where the fiercest bears and boars dwelt in the various reserves (*montes*) throughout his kingdom, as well as how to hunt and kill them. Europeans were not alone in developing customs that protected large, prestigious game species. Many cultures, including those of Australia's Aborigines, protect habitats that abound with game, and restrict consumption of the most delectable foods to older men. The royal game reserve was far from perfect as a mechanism to protect Europe's largest mammals, but it did prolong the existence of the last vestiges of European natural grandeur.

The first extinction to mar the mainland of western Europe since the demise of the muskox 9000 years earlier occurred in 1627, in the Jaktorów forest of Poland. The aurochs was Europe's most magnificent surviving creature. Bulls, which were blackish and much larger than cows, weighed up to 1.5 tonnes, making them, along with the gaur, the largest bovid ever to exist. Cows were reddish-brown and much smaller. Both sexes had an attractive white muzzle and athletic bodies, with deep chests, strong necks and shoulders atop long legs, making them as tall at the shoulder as their body was long. Their enormous horns, up to 80 centimetres long and 20 centimetres in diameter, curved through three orientations: upwards and outwards at the base, then forwards and inwards, and, at the tip, inwards and upwards. The form of the beast, and especially its horns, is evident in many European ice-age depictions.

In Roman times the aurochs was still widely distributed, but by about 1000 CE it had become restricted to a few parts of east-central Europe. By the thirteenth century it is likely that just a single population survived, around Jaktorów, in the Polish province of Masovia. Today, Masovia is the most populous of Poland's provinces, but 700 years ago it was remote and heavily forested. While other large mammals were often hunted by the nobles, the local monarchs, the Piasts, were acutely aware of the value of the aurochs, and reserved the hunting of the creatures for themselves. The penalty for infringement was death.

According to that Boswell of the Polish aurochs, Mieczysław Rokosz:

> The local princes of the Piast dynasty, and later on the kings of Poland, made no concessions of their exclusive right to hunt that animal, not even to the greatest magnates, both ecclesiastical and secular. They themselves never abused the hunting law as far as the aurochs was concerned. Considering the situation of the aurochs in the light of that regale and of the hunting law, the conclusion is offered that the fact of excluding the aurochs from the hunting law and extending to it 'a sacred privilege of immunity' which, according to an old custom, only the king was not obliged to obey, was the major factor which contributed to such a long period of survival of that species. This exceptional and almost personal care of the Polish sovereigns for these animals and their intentional will to save them for posterity caused the prolongation of the period of survival of that magnificent species.[2]

Despite this exceptional protection, by the end of the sixteenth century the aurochs survived only in a small region near the Pisa River. A report by inspectors of the aurochs herd, made in 1564, gives a clue as to why royal protection was not enough:

> In the Jaktorówski and Wislicki primeval forests, we found a herd of about 30 aurochs. Amongst them were 22 mature cows, 3 young aurochs and 5 calves. We could not see any mature

males, because they had disappeared into the forest, but we were told by the old gamekeepers that there are eight of them. Of the cows, one is old and skinny, and will not survive the winter. When we asked the keepers why they are skinny and why they do not increase in number, we were told that other animals kept by village people, horses, cows and so on, feed at places for aurochs, and disturb them.[3]

Being the king's beast was both a blessing and a curse. A blessing because nobody could kill you, but a curse when it came to whether you or the villagers' cows got grass. When feed was short, villagers' self-interest won out, and by 1602 there were just three male and one female aurochs left. In 1620 just one female remained, and when the king's inspector returned to see the aurochs in 1630, he discovered that she had died three years earlier.

The decline to extinction of Europe's wild horses is less well documented. During the Palaeolithic, wild horses had abounded, yet a few thousand years later they had all but vanished from the European central lowlands. In Britain, horses were extinct by 9000 years ago, and 5000 years of horselessness ensued. A similar situation prevailed in Switzerland, where horses became all but extinct 9000 years ago, and remained absent until about 5000 years ago, when the domestic horse arrived.[4] In parts of France and Germany wild horses returned from the brink of extinction between 7500 and 5750 years ago, perhaps the result of forest clearance by humans, which opened more habitat.[5] Yet a different pattern is seen in Iberia, where open habitats persisted naturally, allowing the horse to thrive until about 3500 years ago, during the early bronze age, by which time domestic horses were present.

Herodotus recorded seeing wild horses in what is now the Ukraine, and reports of wild horses in what is now Germany and Denmark continued until the sixteenth century. Wild horses may have survived until the seventeenth century in the part of East Prussia known as 'the great wilderness' (today's Masuria, in Poland). Within a century, however, the great wilderness had been emptied of its horses and only a few captives survived in a zoo set up by Count Zamoyski in south-east Poland, where they persisted until the late eighteenth century.[6] It's possible that wild horses, known as tarpans, survived in southern Russia until the nineteenth century, but these may have been hybrids, with genes from domestic horses. The last tarpan, which bore some resemblance to a domestic horse, died in a Russian zoo in 1909.

Following the tragic loss of the aurochs (and setting aside the wild horse) Europe managed to avoid another extinction for precisely 300 years. The largest wild mammal in Europe was the wisent—bulls occasionally exceed one tonne in weight, with females typically half as heavy. A hybrid between the steppe bison (from which the American buffalo is descended) and the aurochs, the wisent had always been more abundant and widespread than the aurochs, which no doubt favoured its survival for centuries after the aurochs vanished.

Nobody who has spent a moment studying ice-age art could mistake a wisent for any other creature—except perhaps Europe's extinct steppe bison. Their distinctive shape, dominated by massive forequarters, shaggy beard and fringe of fur in the underside of the neck, seems to me to embody the ice age itself. Meeting a wisent face to face, and taking in its distinctive smell, incredible bulk, the steaminess and deep sound of its breathing, conjures a presence most prehistoric.

The wisent is on average a little lighter than the American bison, though it is taller at the shoulder and uniformly brown, with longer horns and tail. Some of these characteristics are seen in the aurochs— one of its ancestors. Dwindling genetic diversity, resulting from small populations and disjunct distributions, reveals that both the wisent and the aurochs were slowly headed towards extinction by about 20,000 years ago.[7] Small, isolated wisent populations survived in the

Ardennes and Vosges mountains in France until the fifteenth century, and in Transylvania until 1790; the very last wisents existed in two small, isolated populations—one in the Caucasus, and the other in the Białowieża Forest in Poland.

The extinction wave finally broke upon the wisent as Europe's human population was thrown into a period of unprecedented butchery. The period 1914–1945 was Europe's darkest hour. After a thousand years of tribal wars, the Europeans, armed with weapons of unimaginable destructive power, set upon each other with horrifying ferocity. All law was set aside, and all care for nature forgotten.

The Białowieża wisents were the legal property of the Polish kings and were strictly protected. But in the turmoil of World War I German soldiers shot 600 for sport, meat and trophies, and by the end of the war just nine remained. Poland was famine-struck in 1920, and the last wild wisent in the country was killed in 1921 by Bartholomeus Szpakowicz, a poacher.[8] Meanwhile, the wisent population in the Caucasus limped on. It is estimated that there were about 500 Caucasian wisent in 1917, but by 1921 only 50 survived, and in 1927 the last three were killed by poachers.[9]

The wisent was not, however, entirely lost. A single Caucasian bull had been taken into captivity, as were fewer than 50 individuals from Białowieża, and the small captive herd was dispersed between various European zoos. It was only the deep sense of loss felt by the Poles that saved the wisent. In 1929 a Bison Restitution Centre was established at Białowieża, and the captive animals were gathered and divided between two breeding groups, one of which was descended from just seven cows, the other coming from twelve ancestors, including the Caucasian bull.

Despite the care taken with herd management, the wisent's genetic diversity has continued to decline, with all males alive today being descend from just two of the five bulls surviving in 1929. Thankfully, the genetic bottleneck appears to be having only a minor deleterious effect on fitness. More than 5000 healthy wisent now live scattered across the Netherlands, Germany and many countries in Eastern Europe. After the closest of shaves with extinction, and barring more human chaos, the future of the wisent seems secure.

CHAPTER 37

Survivors

The next-largest European mammal, after the wisent, is the moose, which in Europe can weigh up to 475 kilograms. Like the wisent, it was in severe trouble a century ago. Long extinct in France, Germany and the Alps, where it had flourished 1000 years earlier, its last stronghold was in Fenno-Scandinavia where, amid the fastness of northern fens, it had survived the butchery that raged to the south. Today a secure, if limited, population continues to thrive in the far north. After the moose comes the red deer, a hardy survivor with flexible food and habitat requirements. Males weigh about 300 kilograms, and females half that. Maturing at two years old, and able to produce a fawn every year, its rapid rate of reproduction has helped it endure intense hunting pressure. Yet, as Europe's human population expanded, even this most resilient of species was pegged back. By the nineteenth century, red deer had become extinct in most of Britain, except Scotland, and where it survived in western Europe, it depended on a degree of protection. When law and order broke down, and hunting laws were disrespected, such as during World War I, its populations suffered. One instance occurred in Germany during the revolutionary years of 1848–49. DNA recovered from antlers collected as hunting trophies by the princes of Neuwied over a period of 200 years reveal a declining genetic diversity of the red deer herd, with a big drop during the 1848 revolution.[1]

The Italians have excelled in preserving vulnerable large mammals in their royal hunting reserves. European ibex, smaller than the other threatened species included here, are a distinctive and vulnerable element in Europe's large mammal fauna.[*] Once widely distributed in alpine habitats across Europe, they reached their nadir in the early twentieth century when just a few hundred survived in what is now Italy's *Gran Paradiso* National Park and the adjoining Maurienne Valley in France. The *Gran Paradiso* was originally the *Riserva Reale* of King Vittorio Emanuele, established in 1821. It is the only reason we still have ibex today.

During its battle for survival, the ibex had to deal not only with poachers and out-of-control soldiers, but with international piracy by— of all people—the Swiss. The ibex had long since been exterminated in Switzerland when, in 1906, the Swiss decided they wanted to restock their Alps. They applied to the Italian authorities for permission to capture some ibex, but were refused. Undeterred, a few wealthy Swiss privately funded a clandestine group that managed to bribe the guards of the *Riserva Reale* and steal almost 100 of the animals. These were transported to the pocket-handkerchief-sized *Peter und Paul* park in St. Gallen canton, where they succumbed to a tuberculosis epidemic.

One hundred years later, in 2006, in a belated act of reparation, the Swiss donated 50 ibex (whose ancestors had been acquired legally) to three Italian protected areas where efforts were being made to build ibex numbers. For all their drama, the Swiss thefts were small beer compared to the ravages inflicted by German soldiers and Italian poachers during and after World War II. Such was the devastation that by 1945, the reserve's guards could find just 416 ibex. With poaching

---

[*] Females weigh 17–34 kilograms, and males 67–117 kilograms.

unabated, the last wild-living ibex looked set to follow the aurochs and wisent into extinction in the wild.

The ibex escaped extinction because of the almost superhuman efforts of Renzo Videsott. He was working in the *Riserva Reale* during World War II, trying to protect the last ibex. But he was also leading a double life as a member of the clandestine, anti-Fascist resistance movement *Giustizia e Libertà*. As he strived to prevent German troops from shooting the last ibex for sport and trophies, his own life often hung in the balance.

Between the end of the war and 1969, Videsott was *Commissario Straordinario* of the *Gran Paradiso*, which was still officially a hunting reserve. He opposed all applications to hunt the ibex, and organised an efficient system of guards, some of them former poachers. In effect, he was commander in a war against poaching, overseeing battles in which people were often killed and wounded. Brave enough to stand up to German soldiers, corrupt Italian politicians and armed poachers, Videsott withstood huge political pressures and considerable personal intimidation in his job and was forced to live under continuous armed guard. Both poachers and guards came from the park's villages, and within each village the guards and poachers were often related, making it difficult for a guard to act against a poacher from his village. But there was fierce competition and hostility between the villages, and Videsott used this to his advantage, posting guards in villages where they had no alliances.*

To avoid detection, the guards would travel by foot across country (not easy in winter in the Alps) and would ambush poachers returning from a kill. But often poachers would bury the ibex in snow, expecting the guards to be alerted by the shot. So the guards sometimes waited

---

* In their old age the guards told Luigi Boitani of their adventures while serving Videsott.

days or weeks in the dead of winter until the poacher returned to
his kill. Due to such actions the ibex survived, and today they are
the pride of the European Alps, with more than 20,000 individuals
distributed from France to Austria, with introduced populations in
Bulgaria and Slovenia.

We are often appalled at the tragic prevalence of wildlife poaching
in Africa, where only a resolute if underfunded band of park guards
works to prevent the total elimination of rhino and elephant.[*] But 70
years ago things were even more desperate in Europe, for Europe had
lost its megafauna, and even its wisent had been driven into extinction
in the wild. Its largest surviving wild creatures were antelope-sized, and
even some of them were being exterminated by the most determined
poaching. The lessons of history should make the world more helpful
to the dozens of unsung African Renzo Videsotts working today. With
a little help, they may succeed in conserving some of Africa's fauna.

While Europe's largest herbivores survived by the good grace of
kings, and often suffered during periods of human upheaval, the opposite was true for Europe's carnivores. They were relentlessly persecuted
almost everywhere, but whenever humans suffered, or chaos reigned,
they thrived. Without doubt, the most hated and feared of all Europe's
carnivores was the wolf. As human populations grew, and with them
the numbers of domestic sheep, goats and cattle, the wolf was persecuted with the utmost determination, and in its history can be read the
fate of Europe's carnivores as a whole.

Charlemagne was a great wolf hater. Between 800 and 813 CE
he established a special corps of wolf hunters known as *la louveterie*,
whose only task was to persecute wolves with hunting, traps or poison.
La louveterie was organised as a military corps, and its salaries paid by
the state. It worked pretty much continuously for more than 1000 years—

---

[*]Organisations such as Parks Africa and the Thin Green Line assist such rangers.

except for a brief break during the French Revolution—which is almost as long as any institution, bar the Catholic Church, has operated in Europe. And it was highly efficient, in 1883 alone accounting for the deaths of at least 1386 wolves.[2] After 1000 years of effort, the *louvetiers* finally put themselves out of business when they killed the last French wolf, in the Alps, at the end of the nineteenth century.

Italy had its own traditional wolf hunters, the *lupari*. They were local farmers who received no prescribed compensation for their work. In a custom called *la questua*, whenever they killed a wolf they would put the body on a donkey and tour the local villages with it, asking for a reward for the service they had done for the community. The *la questua* is probably the reason that wolves were never exterminated from the Apennines. The *lupari* always left a few to ensure they had future income.

By the middle ages wolf persecution had become systemic in Europe. Organised hunts and drives accounted for the extinction of many populations, while overhunting of wolf prey by humans, along with removal of vegetation, made life hard for the survivors. The English got rid of the wolf in the fifteenth century by cutting down most of their forests. Scotland followed with complete eradication through hunting in 1743, as did Ireland in 1770. Persecution continued into the twentieth century. Yugoslavia set up a wolf extermination committee in 1923 that nearly succeeded in its goal—with only a few individuals surviving in the Dinaric Alps. Sweden's last wolves were pursued by snowmobile and systematically shot until the last perished in 1966. In Norway, the last wolf died—again at human hands—in 1973. Had it not been for their long border with Russia, which wolves disrespected, those expert hunters the Finns also might have succeeded in eliminating their wolves.

Despite the persecutions, the wolf had its good times too. When the Black Death ravaged Europe between 1347 and 1353, killing an estimated 30 to 60 per cent of Europe's human population, the wolf was on easy street. In Sweden, for instance, many farms in marginal areas were abandoned to the encroaching forest, and wolf hunting ceased. Consequently, wolf depredations became so extensive that in 1376 the king sent a letter to his subjects explaining that bears and wolves were damaging livestock, and demanding that citizens bring in their skins.[3]

Europe's bears suffered as severely as its wolves, though their persecution was not as systematic. Prior to about 7000 years ago, brown bears were widespread in Europe, their remains being present in 27 per cent of the more than 4000 archaeological deposits examined. But with the growth of agriculture and a warming climate, the human population increased and the bears declined. The temperature increase was detrimental because winter temperatures increased faster than summer ones, and high winter temperatures make it harder for bears to hibernate. Starting in the southwest, Europe's bears began to disappear.[4]

The real crisis did not hit, however, until Roman times. Perhaps the Romans hunted the bears to protect their livestock, or captured and killed them for public entertainment: whatever the case, Europe's bear population fragmented. The brown bears of Scotland were prized by the Romans for their pugnacity, but by 1000 years ago they were extinct. Throughout central and western Europe, brown bears were banished to remote strongholds, with tiny populations holding on in rugged mountain areas in Italy and Spain and a small number in the far north—in Sweden and Finland. The decline continued until the late twentieth century.

Human persecution of the brown bear may have altered its ecology. Elemental analysis of bones has shown that in times past Europe's brown bears were much more carnivorous than they are today. Brown bears that take livestock are hunted down and killed, and it's reasonable to think that this has happened since the dawn of agriculture. Because food preference is at least partially genetically determined, it's easy to see how the very strong selective pressure might have led to the current largely vegetarian population.

Anyone who has met a European brown bear in the wild will have noticed that these great, shaggy beasts that could kill you with the swipe of a paw display abject terror at the sight of a human, and flee at the first opportunity. How different is this behaviour from that of the polar bear, which in the far north has had very limited contact with people, and which, according to the nineteenth-century explorer Adolf Nordenskiöld, approaches man 'in hope of prey, with supple movements, and in a hundred zig-zag bends, in order to conceal the direction he intends to take, and thus keep his prey from being frightened'.[5] Perhaps, before they learned how dangerous humans are, Europe's brown bears also behaved like this.

There are parallels, I think, between the impact of Europeans on brown bears, and the effects of domestication, particularly on dogs. Both sets of selective pressure have altered the behaviour and diet of the beasts in question. Admittedly, Europe's bears still live in the wild, but an argument can be made that the Europeans have domesticated wild Europe itself. It would be well worthwhile assaying the behaviours, diets and reproductive patterns of Europe's wild animals to determine just how greatly human hunting and habitat alteration have changed them.

The catastrophic decline of Europe's large mammals over the past 40,000 years has occurred roughly in order of size. One potent explanation for this is that 'hunters focus on large adult animals (particularly males) to maximise return'.[6] This drives extinction first of the species with the largest body mass and so on down through a size gradient. Within species, this same phenomenon can result in the selection of early maturing dwarfs. 'The Emperor of Exmoor' was a supersized red deer with a magnificent rack of antlers that proved irresistible to hunters. His death, in 2010, tells us a lot about the evolutionary pressure large mammals have been under ever since upright apes populated Europe more than 1.8 million years ago. At 2.75 metres tall and 135 kilograms in weight, the twelve-year-old red deer stag was the largest wild animal in the UK. Even so, he was a runt compared to his ancestors of 12,000 years ago, which would have reached more than twice his weight. The fact that Britain became an island must have contributed to the shrinkage of red deer living there, but the impact of human hunting cannot be discounted as a powerful influence. One study shows that over just ten generations, hunting for large male red deer can cause a decrease in the average body size.[7]

The Emperor was killed during the rut, and may have not passed on his genes before his death (red deer stags do most of their reproducing during the few years they are in their prime). Some months after his demise, the Emperor's head and magnificent antlers mysteriously appeared, mounted on the wall of a local pub. As every fisherman knows, killing an oversized individual can bring prestige as well as meat or monetary profit. I suspect that the same has been true since the stone age, and that indeed some ice-age art is making the same statement as the Emperor's mounted head.

CHAPTER 38

Europe's Global Expansion

After Columbus discovered a sea route to America in 1492, our globe was transformed biologically as well as politically by the great European expansion. By the fifteenth century there were two principal contenders—Europe and China—with a chance of establishing a world empire, and the favourite was China. A unified political entity with a population of 125 million, it was the largest polity on Earth. Europe, in contrast, had a population less than half that of China's, and its states, despite being unified by religion, were perpetually at war with each other.

Both China and the European state of Portugal possessed leaders interested in pushing the boundaries of exploration. In the early fifteenth century the Yongle Emperor instructed his eunuch admiral Zheng He to undertake an epic series of explorations as far afield as Java, Ceylon, Arabia and east Africa, aboard the largest and most advanced ocean-going vessels ever constructed. Chinese junks carried nine sails, had four-tiered decks, were steered using stern-mounted rudders and possessed internal, watertight bulkheads.[1] Guided by the magnetic compass, they had sailed as far as East Africa by the 1420s, carrying hundreds of people, along with those great Chinese inventions: paper money and gunpowder.

The Portuguese Prince Henry the Navigator also devoted his life to exploration; he sponsored a series of voyages down the west coast of Africa. His major breakthrough came with the invention of the caravel—a small, manoeuvrable vessel that made it possible to sail independent of the prevailing wind. By 1418 the Portuguese had discovered and settled Madeira, and by 1427 they had discovered the Azores. Just after Henry died in 1460, the Portuguese pushed down the west African coast as far as Sierra Leone. Although European histories celebrate Henry's efforts, by Chinese standards, they were puny.

Darwin's rule of migration favours larger entities in the evolutionary race, and, given its technological advantages, China was the clear frontrunner. But other factors were against it. The Chinese were not, and never had been, a colonising maritime power. Their battles for expansion and control were fought on land. Zheng He's achievements were thus an anomaly, and soon forgotten. The Europeans, in contrast, had been undertaking maritime colonisations for at least 10,000 years. And they lived around a natural training ground: the *Mare Nostrum*, 'our sea' as the Romans called the Mediterranean. Beginning with the discovery and settlement of Crete 10,500 years ago, the earliest European farmers used ships to reach island after island, in a tradition that thrived right up to Carthaginian times, when Europeans briefly ran out of accessible, habitable islands to colonise. But by the ninth century the process had started again with the Viking discovery and settlement of Iceland, Greenland and, ultimately, North America. During the fifteenth century, Basque fishermen rediscovered Newfoundland, Columbus reached the Caribbean, and the Portuguese sailed to India.

By the time of Prince Henry, the Europeans had one important new tool in their exploration toolbox: the classical worlds of Greece and Rome. Henry could have read Homer and Plato, Plutarch and Strabo, and within fifteen years of his death his successors could read

Herodotus. During the dark ages these texts had been lost to the imaginations of the western Europeans. But now they were learning once again that the world is round, and a very large, exciting and strange place.

When their maritime expansion began in earnest, the Europeans rapidly adapted to the opportunities they encountered in newly discovered lands in ways that extended their traditional ecological niche. Where stratified societies already existed, European colonisation consisted of a sort of social decapitation, in which the ruling class were replaced with European overlords. The conquests of the Aztec and Incan empires and various Indian kingdoms fit this pattern. Where populations were less dense and the ecology suited the Europeans, such as in temperate North America, South Africa and Australia, they followed the time-honoured tradition of settling the new lands as farmers. There were, however, some regions so inhospitable to the Europeans—such as much of equatorial Africa, or so remote—such as New Guinea—that the European presence, where it manifested at all, was fleeting.

In the animal world, there are very few species that have paralleled the European expansion by starting on small landmasses and successfully colonising large areas, but instances do exist. The most striking is the Pacific rat (*Rattus exulans*), a pint-sized rodent similar to the black rat, but weighing less than half as much. It originated on the tropical island of Flores (which has an area of only 13,500 square kilometres) in the Indonesian archipelago, and it remained there until about 4000 years ago.[2] When voyaging ancestral Polynesians touched land there, the rat boarded their canoes. Today the Pacific rat is spread from Myanmar to New Zealand, and from Easter Island to Hawaii, making it one of the most widely distributed small mammals on Earth. How and why was this particular animal so successful at spreading from its homeland? After all, the Indonesian archipelago (and indeed the world) abounds with rats. Not coincidentally, Flores was also home

to the pint-sized humanoid, known as the hobbit (*Homo floresiensis*).
Weighing a third as much as an adult human, it stood only as high as
a three-year-old. The hobbit's ancestors may have reached Flores two
million years ago, so there was plenty of time for the diminutive Pacific
rat to form an ecological relationship with it.* The hobbit became
extinct 50,000 years ago, about the time the first humans arrived on
Flores, but the Pacific rat lived on. Perhaps the small rodent had discovered that the campsites of upright apes are congenial living places. To
put it in ecological terms, the Pacific rat may have been pre-adapted to
spreading into human-modified habitats through its long association
with the hobbit. So, the Pacific rat and the Europeans appear to break
Darwin's rule of migration for very different reasons: the Pacific rat
was pre-adapted to human-created habitats, while the Europeans were
pre-adapted to a colonial lifestyle because they are a maritime people
who originated at the crossroads of the world.

---

* They were certainly present by 800,000 years ago.

HUMAN MOVEMENTS INTO AND OUT OF EUROPE

38,000 years ago	Europe is colonised by humans from Africa, forming a hybrid human–Neanderthal population.
14,000 years ago	Western Europe is colonised by people from the East.
10,500 years ago	Europe is colonised by agriculturalists from western Asia.
5500 years ago	Europe is colonised by horse-riding people from central Asia.
2300 years ago	Western Asia, parts of India and north Africa are colonised by Alexander the Great.
From 2200 years ago until seventeenth century CE	Steppe nomads, Arabs and Turkic peoples invade Europe.
1000 CE	Norse colonisation of Newfoundland.
1500 CE	Beginning of European colonisation of most of the globe.
Mid-twentieth century	European decolonisation of most of the globe.

CHAPTER 39

New Europeans

Myriad wild creatures have made a home in Europe after humans imported them, but not a single species imported by the Romans ever established itself. Surely no people imported a greater diversity of creatures into Europe, so this fact is as astonishing as is the realisation that the Romans didn't add a single species to the domestic menagerie. The Romans did, however, spread species within Europe, including bringing fallow deer and brown hare to Britain. The black rat spread wherever they settled, and so tied was it to Roman habitats that it became extinct in Britain following the Roman withdrawal, only to reappear in Norman times.* Thereafter the black rat prospered—until the arrival of the larger brown rat (also known as the sewer rat or Norway rat), which reached Britain at about the time of the Hanoverian succession in the eighteenth century. Renowned naturalist 'Squire' Charles Waterton called it the 'Hanoverian rat': a devout Catholic, and lover of British wildlife, he viewed its ravages

---

*The black rat originated in southeast Asia. There is evidence of black rats in Europe and the Levant earlier than Roman times, but it only becomes widespread in the wake of the Romans. The fallow deer introduced to Britain by the Romans seem to have become extinct. The species was reintroduced during Norman times.

and the influence of the German-speaking monarchs as equally pernicious.[*]

The Romans may also have had a hand in the spread of that most noble game bird, the ringneck pheasant. Originally Asian, the pheasant had spread as far west as Greece by at least the fifth century BCE, and Pliny mentions its presence in Italy in the first century CE. Its introduction to Britain, at least, may be down to the Romans: pheasant bones have been recovered from at least eight Roman archaeological sites in Britain. The possibility exists, however, that these birds were not raised in Britain, but were brought in from elsewhere.[1]

Like the black rat, the pheasant seems to have disappeared for a period after the Romans left Britain. The first written record in Britain dates from the eleventh century, when King Harold offered the canons of Waltham Abbey a 'commons' pheasant. The first clearly wild populations (which were protected by royal decree) date to the fifteenth century.[2] The current British population is a hybrid: as one British game farmer put it, 'just about every species and subspecies has been crossed by now.'[3]

Some 8000 years after the last members of the elephant family vanished from Europe, pachyderms made an unexpected and terrifying return. During the Second Punic War, 218–201 BCE, Hannibal trekked 37 war elephants across the Alps from what is now Spain into Italy. Just what species they were is hotly debated. A coin minted at the time shows what is clearly an African elephant. But the only creature to survive the war—Hannibal's own mount—was called Surus, meaning 'the Syrian', suggesting an Asiatic origin.

---

[*] Waterton was a genuine eccentric who delighted in dressing as a scarecrow and sitting in trees, pretending to be a dog and biting the legs of his guests or tickling them with a coal brush. He was much feared on his estate because of his enthusiasm for treating ailments that appeared among the tenants. Bleeding was invariably part of the cure.

Hannibal's elephants were almost certainly not straight-tusked types, as the coin shows a creature with curved tusks. Moreover, by Roman times the straight-tusked elephants were restricted to equatorial Africa, which is very distant from Carthage. Hannibal's elephants were possibly sourced from a now-extinct population of African elephants from the Atlas Mountains, which consisted of rather small individuals. But if some at least were Asian, they may have been derived from Indian war elephants captured by the Ptolemies of Egypt during their campaigns in Syria. Wherever they were from, they clearly took to life in Europe, surviving the alpine snows and thriving sufficiently to strike terror into the Roman legions.

Sometime after Rome was sacked by the Visigoths in 410 CE, the highway to Africa re-opened. This time, however, instead of being a land bridge it consisted of Moorish vessels. The Moors, who settled many parts of southern Europe in the eighth century, proved to be enthusiastic naturalisers. They are strongly suspected of, or were clearly responsible for, the introduction of at least four important mammal species into Europe: the Barbary macaque, porcupine, genet and mongoose. The genet and mongoose arrived sometime after 500 CE. The genet is a handsomely spotted member of the ferret family which had been brought into captivity by the Moors (it is still occasionally kept as a pet in north Africa), but some escaped and the creatures now abound in southwestern Iberia, where the Egyptian mongoose too has found a home.

The Barbary macaque had been resident in Europe for millions of years, extending as far north as Germany during warm periods, but by 30,000 years ago it had become extinct in its last European stronghold in Iberia. It survived, however, in North Africa, and from there was re-introduced to Gibraltar at around the same time that the genet and mongoose arrived, although the first written record of the species dates only to the 1600s. The Barbary macaque would have become extinct on

Gibraltar were it not for a peculiar belief by the British that they would hold the rock only for as long as the macaques remained. The British have occupied Gibraltar since 1713, but by 1913 only ten macaques remained. Some years later, to prevent total extinction, the governor of Gibraltar, Sir Alexander Godley, brought eight young females from North Africa, and the British army assumed responsibility for them. At the outbreak of World War II only seven apes remained on 'the rock', and Churchill commanded that five females be brought in from Morocco, with a directive that the population be maintained at 24. By 1967, when Spain looked set to claim back Gibraltar, the macaques were again in decline. Concerned at the severe sex ratio imbalances in some groups, the permanent under-secretary at the Commonwealth Relations Office sent a telegram, reminiscent of something from a *Carry On* movie, to Gibraltar's governor:

> We are a little perturbed about the apes…As we see it, at first glance there seems at least some chance of lesbianism, or sodomy, or rapes…the Queen's Gate lads, one fears, may become a bunch of queers…So can you plan migration?[4]

Today the 230-odd Barbary macaques on Gibraltar are the responsibility of the Gibraltar Ornithological and Natural History Society. They remain Europe's only wild-living monkeys.

The crested porcupine also appears to have arrived in Europe sometime after the sack of Rome in 410 CE.[5] At up to 27 kilograms in weight, it is a very large rodent. Fossils indicate that porcupines once inhabited Italy and other parts of Europe, possibly until about 10,000 years ago, but they may have been a different species. Today, the porcupine is restricted within Europe to peninsular Italy, though it is spreading steadily north.

Were the Moors trying to 'Africanise' Europe with their introductions of mammals? The Moorish homeland was outside Europe, and some Moors were desperately homesick. Abd al-Rahman I's ode to a palm tree which was, 'like myself, living in the farthest corner of the earth' sparked a major theme in Andalusian poetry.[6] After the fall of Granada in 1492, and the expulsion of the Moors from Europe, there was a lull in animal introductions which, with few exceptions, lasted until the European age of empire.

One important exception is the common carp (*Cyprinus carpio*), an oversized minnow originally found in the lower Danube and other rivers flowing into the Black Sea. It arrived in western Europe around 1000 CE courtesy of monks, who raised the fish in ponds to help people observe religious fast days, when fish but not red meat could be eaten. Within a few hundred years, carp farming had become big business, and the carp had transformed into a wild-living denizen of many European waterways.[7]

CHAPTER 40

Animals of Empire

The next big wave of migrants would arrive from across the Atlantic. Ever since the Vikings had planted settlements on the coast of Labrador in the tenth century, Europeans had been incrementally opening a connection with the New World that had last existed before *la grande coupure*, 34 million years ago. The establishment of the new Columbian maritime highway in 1492 meant that Europe again became the lands at the crossroads of the world, for it lay at the confluence of a global trade network encompassing Asia, Africa *and* the New World. In the sixteenth century Montaigne described the catastrophe that came with the European expansion:

> So many goodly cities ransacked and razed; so many nations
> destroyed and made desolate; so infinite millions of harmless
> people of all sexes, states, and ages, massacred, ravaged, and put
> to the sword; and the richest, the fairest, and the best part of the
> world topsy-turvied, ruined, and defaced for the traffic of pearls
> and pepper.[1]

But the impact was felt within Europe as well, for such was the flood of plants and animals into Europe that its ecosystems would be

thoroughly topsy-turvied. And some European species would be driven
to the brink of extinction.

One of the most pervasive products of the Columbian highway
was yet another influential European hybrid—the London plane tree.
Ancestral plane trees, you may recall, flourished in Europe 85 million
years ago, during the age of dinosaurs. As such, they are one of Europe's
living fossils. A family of just one genus and a dozen species, the plane
tree's nearest living relatives are the proteas and banksias. The hybrid
London plane, which lines city streets worldwide, is by far the most
familiar species. Referred to by the botanical writer Thomas Pakenham
as that 'mysterious bastard', its precise origins remain obscure. But
among its parent species are the oriental plane tree (*Platanus orientalis*)
and the American sycamore (*Platanus occidentalis*).[2]

The oriental plane tree has a most strange history. It is native to
southeastern Europe and the Middle East; expanded into western
Europe at about the same time that the grapevine, olive, chestnut and
walnut were brought west by early farmers. But while these species are
important food plants, plane trees produce nothing useable, not even
timber. Perhaps Neolithic people enjoyed their beauty, and the shade
they offered in summer.

The parent species (oriental plane and American sycamore) occur
naturally in eastern Europe and eastern North America respectively,
and have probably been separated at least since the Arcto-Tertiary
Geoflora was disrupted by the onset of the ice ages 2.6 million years
ago.[3] The London plane first appeared through hybridisation in the
seventeenth century, and soon became valued in heavily polluted, early
industrial Europe, both because it tolerated air pollution, and because
its bark fell away in flakes, self-cleansing the trunk in the sooty air.

The Romans valued the sweet chestnut for its shiny brown fruit,
and so spread it widely in Europe. And, since the age of empire, garden
escapees have been enriching European forests. Among the most

able colonisers are those whose seeds are spread by birds: this is why Asian windmill palms and camphor laurels sprout in the foothills of the European Alps. Even the Australian eucalypts are managing to establish forests in the lands surrounding the Mediterranean. In fact, Australia's most widespread eucalypt, the river red gum, was named from specimens growing in the grounds of the Camaldoli monastery, west of Florence, some time prior to 1832.

For centuries the trans-Atlantic traffic of animals had been one-way—from Europe to the Americas. It's only during the last 200 years that American creatures have begun to establish in Europe. The fur trade has a lot to answer for. In the 1920s American mink escaped from fur farms and became established in the wild. More recently they have been joined by American mink released by animal rights activists. The species has become established throughout much of Europe, and it has displaced the European mink, which is now critically endangered.

American beavers were released into the wild in Finland in 1937. At the time it was wrongly thought that American and European beavers were the same species, and the American beavers were brought to Finland as part of a reintroduction program and as a source of fur. But they proved competitively superior to the natives, and by 1999 it was estimated that 90 per cent of beavers in Finland were the American species. Eradication efforts to preserve the native species are ongoing.

The muskrat is a medium-sized aquatic rodent from North America that is now found throughout much of temperate Eurasia. It, too, was introduced a century ago as a source of fur, and quickly escaped. Muskrats damage dykes, levees and crops, but almost all efforts at eradication, or even control, have been given up as hopeless. I watched one swimming in the new European 'wilderness' at Oostvaardersplassen in the Netherlands, pleased at least that there was no European version of the muskrat for them to outcompete.

The coypu is a large South American rodent with the habits and appearance of an oversized muskrat. Coypus were first brought to Europe in the 1880s for their fur, and like the mink and muskrat, soon escaped. Because of the damage they do to semiaquatic vegetation, eradication programs were instigated in many countries, including the UK, where, after an expensive and arduous campaign, eradication was finally achieved in 1989. However, coypus remain widely distributed throughout continental Europe, and are likely to increase in number and distribution in the warming climate.

The Allied victory in Europe in 1945, in which the Americans played a central role, ushered in its own influx of American invaders. Perhaps this was because, for a few decades following the end of World War II, it seemed that America could do no wrong. With tens of thousands of American troops stationed across the continent, a motley array of American military mascots and pets were poised to find their own homes in Europe. The North American grey squirrel had already been introduced into Britain, in 1876, and today it dominates many areas of England, where it displaces the native red squirrel. In Ireland, grey squirrels drove out red squirrels for decades following their introduction in 1911. But as the native pine martens recovered from human persecution, grey squirrels began to vanish.[4] It may be that grey squirrels lack defences against pine martens, while red squirrels—which had evolved with the predators—are better able to evade them. Perhaps in future a balance will be struck in which pine martens, red squirrels and a much-reduced population of grey squirrels coexist.

Grey squirrels did not arrive in continental Europe until 1948, when two pairs were released at Stupinigi, near Turin. In 1966, five more were imported and released at the Villa Gropallo near Genoa. Finally, in 1994 a third introduction was made at Trecante, again in Italy. The Stupinigi population flourished, and by 1997 it had spread across 380 square kilometres. Today, for various reasons (including a rule set by the Bern Convention, which is the most important treaty on wild species conservation in Europe) the Italians are attempting to eradicate this American intruder.[5]

In the 1950s, the North American cottontail rabbit (*Sylvilagus floridanus*) was also introduced and has thrived, particularly in parts of Italy.[6] How it is interacting with native rabbits and hares is not known. There are now populations of white-tailed deer—another American—in several countries, including the Czech Republic. And in the aftermath of World War II the raccoon became firmly established. Despite their previous enmity, Americans, Russians and Germans (who know it as *Waschbär*) have all done their bit to aid the raccoon's colonisation of Europe.

The rise of the raccoon began in April 1934, when a soft-hearted German chicken farmer beseeched a forester in the north of Hesse to release his pet raccoons into a nearby forest. Despite lacking official approval, the forester obliged, and today the area around Kassel in Hesse has one of the densest raccoon populations of Earth—with between 50 and 150 per square kilometre. A second release—this time of 25 raccoons—occurred in 1945 when an air strike damaged a fur farm at Wolfshagen, east of Berlin. By 2012 it was estimated that more than a million raccoons roamed Germany.[7]

In 1958 the Russians released 1240 raccoons across the Soviet Union, to be hunted for their fur. As a result, the Caucasus is today riddled with the masked beasts. In 1966 American air force personnel stationed in northern France released some mascot raccoons, creating a new plague. With the end of the American century, Europeans are looking at many animal immigrants with a jaundiced eye, and some pests are being exterminated. But others look set to become permanent members of a new European fauna.

A few Asian species have made their way into Europe, including the east Asian Sika deer, a near relative of the red deer. It has populated much of Europe and begun to hybridise with red deer, so is considered by some to be a danger to the native species: yet Europe's rich history of hybridisation should warn against simplistic thinking. A few muntjac, or barking deer, escaped from Woburn Abbey in England in 1925, and their descendants now flourish in England and Wales. Tiny deer with simple horns, and long canines in the males, they are reminiscent of the deer that abounded in Europe more than 10 million years ago.

The raccoon dog is yet another Asian species that has found a home in Europe.* More than 10,000 raccoon dogs were released into the wild at various locations in the Soviet Union between 1928 and 1958, as a source of fur. They were first noticed in Poland in 1955 and in East Germany in 1961. They have now reached central Norway and are expanding into Central Europe. Among the many small creatures they prey on are frogs and toads, which makes me worry about the fate of Europe's most ancient survivors, the midwife toad family. The Danes, at least, have had enough of the raccoon dog, and are exterminating it.

Marsupials may seem like unlikely invaders. Europe's last native marsupials were rather like American opossums, and their lineage died out around 40 million years ago. But the descendants of wallabies that escaped from a zoo in about 2000 thrive in forest west of Paris The French seem to like them, the Mayor of Emance saying that they 'have been part of our daily life for twenty years'.[8] The species has made an even greater success of it in Britain. In the 1970s red-necked wallabies escaped from various wildlife parks, and populations now roam wild on the island of Inchconnachan on Loch Lomond, and in Buckinghamshire and Bedfordshire. A successful wild mob can be found in the Curraghs, a wetland nature reserve on the Isle of Man.

Just one amphibian—the African clawed frog—has become an invasive species in Europe. This tongueless, toothless and entirely

---

* Although superficially similar to raccoons, raccoon dogs are canids.

aquatic African creature is among the oddest of amphibians, renowned for its ugliness and gluttony. It would surely never have left African shores but for the fact it is very easy to keep alive in the laboratory, and so is ideal as an experimental subject. It was the first creature ever to be cloned, and the first frog in space (in 1992 a few were taken on the space shuttle *Endeavour*).

But more than anything else, the species owes its spread to a strange phenomenon discovered in 1930 by the biologist Lancelot Hogben. Heavens knows what led him to do it, but Hogben found that if you injected an African clawed frog with the urine of a pregnant woman, within hours it would lay eggs. Before chemical pregnancy test kits became available in the 1960s, African clawed frogs were kept in labs and hospitals worldwide for confirming pregnancies. Many escaped, or were released, including the founders of a population that became established in South Wales.

It is curious to reflect that the Pipidae (the family to which the African clawed frog belongs) is thought to be closely related to that ancient and extinct European amphibian family, the Palaeobatrachidae. It certainly is similar in its ecology. Perhaps we should overlook its grotesqueries and consider the clawed frogs in Wales as ecological replacements, to some extent, for the venerable European palaeobatrachids.

Last century, Europe's fresh waters were colonised by a large number of invasive species, including the red-eared slider (a terrapin), five kinds of freshwater crayfish, the pumpkin-seed fish, rainbow trout, black bullhead catfish and largemouth bass. All are from North America, though it must be said that the killer shrimp (originally from the region of the Black Sea) has also colonised western Europe.[*] Between 2000 and 2012, California and Louisiana exported more than

---

[*] North America has much larger freshwater habitats than Europe.

48 million red-eared sliders around the world.[9] Unsurprisingly, the creature is now listed as the world's worst invasive species.

After a gap of many millions of years parrots are again winging their way across European skies. Since the 1960s, the rose-ringed parakeet, whose original distribution covers large areas of Africa and south Asia as far as the Himalayas (which has inured it to the cold), has been seen in downtown Rome and the suburbs of London, sometimes perched comfortably in gum trees. The monk parrot, a native of Argentina and adjacent countries, is cold tolerant. It was first seen in the wild in Europe around 1985 and has become established across a vast area. Authorities in the UK have recently become alarmed at its success and may act against it. But new invasive birds continue to arrive, including the Indian house crow, seen at the Hook of Holland in 1998 after stowing away on a ship.* I have also been told that a small population of white cockatoos has become established on the Côte d'Azur. I love cockatoos, despite their propensity to destroy houses by tearing apart wooden windows, doors and lead flashing. Were I a European, I would act against them—more in sadness than in anger—before it's too late.

---

*Four kinds of goose have become invasive in Europe—the Egyptian, bar-headed, Canada and swan goose.

CHAPTER 41

Europe's Bewolfing

Nature abhors a vacuum, and given half a chance it fights back against extinctions. Almost 10,000 years after the lion and striped hyena arrived in Europe, another carnivore is making its way west, all alone and without any conservation support. Until half a century ago the golden jackal occurred only east of the Bosphorus, in Turkey. Somehow, a few managed to enter Greece and the lower Balkans. The most recent sightings (and killings) occurred in Estonia, France and the Netherlands. It seems that soon golden jackals will be strolling Europe's Atlantic shore.

Is this blitzkrieg-like spread the result of a low density of wolves in Europe? The rise of the golden jackal and eclipse of the wolf may be no more than coincidence. For most of the Pleistocene, Europe was home to both a wolf-sized and a jackal-sized canid, the jackal-sized creature being restricted to the Mediterranean region before it became extinct about 300,000 years ago. The golden jackal may be filling the ecological role of its extinct relative. In any case, the golden jackal is an important new meso-carnivore in Europe and it is here to stay.

The arrival of the golden jackal comes at a unique time in European history. After millennia of war, starvation and relentless expansion of the human population, a new prosperity arrived in the decades of peace following World War II. Europe's human population began to stabilise, and to concentrate in the cities and coastal plains. Villages in the more remote and hard-scrabble regions, including some mountain areas, are being abandoned, and nature is starting to creep back. But this time there are no royal decrees demanding renewed efforts to persecute wolves and other wild creatures. These animals are now seen as curiosities to be tolerated, or even cautiously welcomed. Seals cavort at Canary Wharf in London, wolves are seen in the Netherlands, and wild boar wander the streets of Rome. The ecology of Europe has, in just a couple of generations, shifted so dramatically as to usher in a bewolfing (and perhaps a bewitching) of the continent.

By the 1960s the wolf was on the brink of extinction in Europe. Only in Romania was it present in any numbers. But by 1978 wolves were once more in Sweden, the result of a pair that travelled from Finland. The Swedish population really took off when another migrant arrived with a batch of fresh genes. As of 2017, there are more than 430 wolves in Sweden and Norway. Norway aims to maintain a national population target of four to six litters per year and is trying to restrict wolves to a small area along the border with Sweden.

South of Scandinavia, wolves are growing in number almost everywhere. Some of the increasing populations, such as that in France, face intense opposition from farmers. But on the whole, the expansion is, so far at least, uncontroversial. In Germany in 2000 there was just one wolf pack. Now there are now more than 50, and nobody seems to mind. The same attitude prevails in Denmark, where the first wolf litter for several centuries was born in 2017.

In early 2018 a wolf was seen in the Flanders region of Belgium, the first in more than a century. Belgium is the last continental European country to be re-colonised by wild wolves, so, at a national level at least, the bewolfing of Europe is complete. Environmental attitudes, the legal protection given to wildlife by European laws, the increasing densities of deer and wild boar near cities, and the sudden depopulation of mountain and hilly areas, have all aided the wolf's expansion. There are now more wolves in Europe than in the United States, including Alaska!

The bewolfing of Europe is bringing wolves and humans into a proximity not seen since the stone age, and some escalation in wolf–human conflict seems inevitable. With animal liberation movements on the rise, some will call for saving the life of every wolf. Others will seek a compromise between wolf and human needs. As they expand, wild-living wolves are encountering the descendants of wolves who threw their lot in with us 30,000 years ago. Since that time the descendants of those human-loving wolves have been shaped by intensive evolutionary pressure into dogs. And today feral dogs far outnumber wolves. Romania, for example, has 150,000 feral dogs and just 2500 wolves, while Italy has some 800,000 free-ranging dogs, and about 1500 wolves.

The ecologies of wolves and dogs have diverged in interesting ways. Wolves eat deer and other large prey, but after millennia scavenging around our campsites, dogs have learned to eat almost anything, and will kill anything from mice to wisent, with free-ranging dogs forming packs to hunt large mammals. But while hungry dogs may evoke our sympathy, we are more likely to shoot a hungry wolf.

Wolves and dogs can mate and produce fertile offspring. Indeed, they have been hybridising for a very long time, as is evidenced by the laika, a wolf-like dog that even today accompanies various Siberian peoples.[1] Wildlife managers often try to eliminate dog–wolf hybrids, because they fear that hybrids will eventually replace the wolves. But more thought may be needed on this issue. Hybrids are such an important part of European evolution that an argument could be mounted that a hybrid species is more appropriate to a continent so profoundly modified by people. Such hybrids may in any case be a natural evolutionary outcome of wolves living close to humans. Should we accept them as long as they fulfil the same ecological functions of the true wild wolves? The moral question as to whether we *Homo sapiens–H. stupidus* hybrids should allow feral *Canis lupus familiaris* and *Canis lupus lupus* to hybridise is complex, to say the least. In trying to control evolution by preventing hybridisation, we may be acting in potentially dangerous and destabilising ways.

*

In 2004, an Italian brown bear wandered into Germany. JJ1, or Bruno, as he became known, was the first brown bear seen in Germany since 1838. In a country whose capital has a bear on its city seal, you might think that the return of the creature would be celebrated. But on 26 June 2006, just two years after his arrival, Bruno was stalked and shot dead—on Rotwand Mountain in Bavaria. Many Germans were, in fact, delighted by the return of the brown bear, but Bruno came from a problem family. His sad story began years earlier, when ten Slovenian bears were released in the central Italian Alps. Among them were JJ1's parents, Jurka and Joze.

It appears that Bruno's mother Jurka was something of a throw-back to her carnivorous ancestors of thousands of years ago, and her meat lust was inherited by her offspring. By the time of Bruno's death, he had accounted for 33 sheep, four domestic rabbits, one guinea pig, some hens and a couple of goats. As Edmund Stoiber, the president of Bavaria, put it, Bruno was a *Problembär*. And so, in a process begun by our ancestors thousands of years ago, Bruno and his brother were removed from the gene pool, in an effort to ensure that future generations of brown bears will be more sedentary—and vegetarian.

Many people worry about the taming of bears for performance on street corners and in circuses, but few realise how deeply we have altered the ecology of wild bears. Over thousands of years, we have created a more fearful and tractable species—which is, from an ecological perspective, a miniaturised version of the vegan cave bear, and able to survive in today's densely populated Europe.

The people of northern Italy like their wild bears, but in 1999 the local bear population of the province of Trento was down to just two animals, which had no chance of reproducing. So the Trentinos imported 10 bears from Slovenia. The program was a great success, and today there are about 60 bears in the region. But it has not all been smooth sailing. Recently, a female bear was killed by the authorities after she attacked a hiker. About half of the problems with the Trentino bears, incidentally, stem from members of Jurka's and Joze's family, so Jurka, the mother, has been brought into captivity. There can be little doubt that conflict will increase as bear numbers grow. But so far, at least in Italy, both human and bear behaviour is leading to a largely peaceful coexistence.

Elsewhere in Europe bear populations are also recovering. Due to careful conservation, the few bears surviving in Sweden have blossomed over 50 years into a healthy population of 3000 or more, and two tiny populations in northern Spain are increasing after many years of lingering in critically low numbers. But the population in the Abruzzo region, just two hours drive from Rome, is unable to expand. Because of a lack of habitat, it remains frozen at 50–60, making it vulnerable to inbreeding and extinction.

In eastern Europe the brown bear is still present in large numbers in many countries including Croatia, Slovenia, Bulgaria and Greece. But to see a truly thriving population you need to go to Romania, where more than 3000 still roam—thanks to the bear-loving despot Nicolae Ceauşescu who, in a resurrection of the *caccia medieval*, reserved to himself the right to kill bears. So abundant are they that some have taken to rifling through rubbish bins on the outskirts of Braşov, one of the country's largest cities. Today, Europe's bear population overall is in good shape, in larger number than those in the lower 48 states of the USA (where they are known as grizzlies).

The Iberian lynx is the largest carnivore unique to Europe. In the stone age it was widespread in southern Europe, and in historic times it roamed throughout the Iberian Peninsula and into southern France. But half a century ago it began a steep decline, due to a reduction of its prey (rabbits), accidents with cars, habitat loss and illegal hunting. By the dawn of the twenty-first century it had been reduced to just 100 individuals, of which only 25 were breeding females, clinging on in two populations in the Montes de Toledo and the Sierra Morena. A massive captive-breeding program, supported by the European Union and costing €100 million, has dragged the Iberian lynx back from the brink of extinction, and today there are more than 500. Its recovery is one of the greatest conservation successes ever seen in Europe.

The Eurasian lynx is a larger feline that once coexisted with the Iberian lynx. Its distribution, however, was far more extensive, covering most of Europe. By the early twentieth century its last refuges were in Scandinavia, the Baltic states, the Carpathians in Romania (which are *the* European heaven for large carnivores) and the Dinaric Alps in Bosnia and Herzegovina. Between 1972 and 1975 eight wild lynx from the Carpathians were released in the Swiss Jura Mountains, where there are now more than 100, and some have been relocated to the eastern canton of St. Gallen. More reintroductions followed elsewhere in Europe, and now there is even talk of re-introducing the lynx into Scotland.

For centuries seal hunting in Europe was relentless, and many populations were confined to breeding in caves. But people would pursue and slaughter them even there. Irishman Thomas Ó'Crohan left an account of one such hunt, which took place on Great Blasket Island in the late nineteenth century:

> The cave was…a very dangerous place, for there was always a strong swell around it, and it's a long swim into it, and you have to swim sidelong…There was a strong suck of swell running. Often and again the mouth of the hole would fill up completely,

so that you'd despair of ever seeing again anybody who happened
to be inside…

The Captain of the boat said, 'Well, what did we come here
for? Isn't anybody ready to have a go at that hole'? It was my
uncle who gave him his answer: 'I'll go in', said he 'if another
man will come with me.' Another man in the boat answered him
'I'll go in with you,' says he. He was a man who stood in need of
bit of seal meat, for he spent most of his life on short commons…

Ó'Crohan's uncle couldn't swim, but he and the other man went
in with a rope held between their teeth, and matches and candles
under their hats. After a tremendous struggle they managed to kill all
eight seals sheltering in the cave. 'It's odd the way the world changes,'
Ó'Crohan wrote much later, in the 1920s. 'Nobody would put a bit
of seal meat down his throat today…yet in those days it was a great
resource to the people.'[2] When hunting ceased, the grey and harbour
seals recovered. Today Great Blasket Island is abandoned, and hundreds
of grey seals breed on its beaches.

But not all of Europe's seals have done so well. Only about 700 individuals of the Mediterranean monk seal survive, across four populations.
It is an ancient lineage: fossils dating to about six million years ago have
been found in Australia. Until the eighteenth century it reproduced on
beaches, but now it uses only inaccessible caves. Ongoing harassment,
critically small population size, and ocean pollution, continue to jeop-
ardise its future.

Much has been done to restore Europe's raptors. Following reintroduction, red kites once again fly in English skies, white-tailed sea-eagles
soar in Scottish skies, and the lammergeier can be seen in the Parc
Mercantour in the Maritime Alps. Some raptors are even increasing their
range without help, including the sea-eagles of the Oostvaardersplassen,
where a pair self-established in 2006, and have bred annually ever since.

Vultures are also recovering, though not without considerable assistance. A program in the Rhodope Mountains, between Bulgaria and Greece, aims to protect black and griffon vultures. These magnificent birds, which are among the largest of all flying creatures, are threatened by farmers who leave poisoned carcasses to kill predators. Teams with specially trained dogs track the carcasses and try to remove them before the vultures consume them. There are also many griffon vultures in the Balkans, and a new colony in Italy's Abruzzo. Today, some vultures tagged in the Balkans are seen in Gargano and up to Abruzzo, so the populations are joining up. But if Europe is to recover its full suite and population density of large raptors and scavengers, some provision will have to be made for leaving carcasses of domesticated beasts in the field, a practice currently strictly forbidden by the EU, even in nature reserves.

Europe's populations of carnivores, larger herbivores and scavengers are now healthier than they have been for at least 500 years. Despite its human population of 741 million, Europe is once again becoming a wild and environmentally exciting place. But as some of ancient Europe's wild beasts are being resurrected, the familiar 'wild' Europe of hedgerow and field, celebrated in the works of Beatrix Potter, is in eclipse.

CHAPTER 42

Europe's Silent Spring

Europe was the first region on Earth to industrialise, and the first in modern times to experience a massive surge in its population. It was also the first to enter the demographic transition (in which both birth and death rates plunged, allowing the population to stabilise and in some cases decline). In much of Europe the population is now maintained by migration or is falling. These profound changes have been accompanied by the development of a new agricultural economy that has displaced human labour with machines and has seen an intensification of agriculture on the best soils, as has occurred on almost every other continent.

Europe's nineteenth-century agricultural landscapes were the result of human influence over thousands of years, and it's from this ecology that the natural Europe of childhood storybooks is drawn—a landscape of hedgerows, spinneys and wild riverbanks—tiny semi-natural areas that managed to survive the intense application of human labour. It is a Europe of small creatures—of fieldmice, voles, sparrows and toads—that have adapted over thousands of years to living in a human-manipulated landscape. And elements of that landscape had not altered greatly for millennia—until the late twentieth century. The loss of spinney and hedgerow is a great blow to many Europeans, for it is the loss of a dreaming tied to the potent themes of childhood idylls and freedom. But if we wish to maintain them, someone must be willing to work the small fields and hedgerows of a Beatrix Potter-world, with the skills and willingness to live as characters in a Thomas Hardy novel.

The great changes that banished hedgerows sprang from new technologies, and Europe's determination to feed itself, which initiated a process we might call industrial-driven decline. Industrial agriculture requires scale, so the hedgerows were ripped out in favour of larger fields enclosed with wire fencing. And efficient agricultural practices began utilising the small, rough corners of farms that had once been wildlife havens. Then began a wholesale drenching of the landscape with agricultural chemicals—fertilisers, herbicides and pesticides—which proved fatal to many smaller beasts.

Butterflies and even ants are among the victims, though nowhere is the decline more evident than among Europe's birds. A group of researchers has tracked the fortunes of 144 European bird species for 30 years. Using Birdlife International data, they estimated that there were 421 million fewer birds in Europe in 2009 than there had been in 1980.[1] As might be predicted, the greatest losses were among farmland species. The study did, however, reveal large increases in the populations of some rare types, probably as a result of the growth of wild lands in remote areas, and high levels of conservation effort. Extremely large losses have occurred in agricultural landscapes in Germany, where it's estimated that 300 million breeding pairs of smaller birds disappeared between 1980 and 2010—a decline of 57 per cent. Among those hardest hit is the skylark, whose song, once ubiquitous, is now rarely heard. But even the most abundant of birds, such as the song lark, house sparrow (self-introduced to Europe 10,000 years ago) and starlings, are being severely affected.[2]

From the beautiful Corsican swallowtail—one of Europe's most lovely butterflies—through to obscure parasitic ants, many insects are now endangered. The black-backed meadow ant is probably extinct in the UK, while the red-barbed ant, narrow-headed ant and the black bog ant are all endangered due to industrial agriculture. More worrisome, very large declines in the volumes of insects have been recorded, even in nature reserves. The use of pesticides and herbicides is having a massive, yet hidden, impact that is striking at the base of food chains.[3]

The European Union's agricultural regulations have been unable to combat the threats. As one Spanish biologist put it:

> Despite previous reforms, the Common Agricultural Policy (CAP) largely continues to support a resource-intensive and high-impact agricultural model which is not fit for today's societal and environmental challenges.[4]

One assessment marked just 16 per cent of habitats and 23 per cent of species favourably.[5] Surely if there is one thing that all Europeans could agree on it is the need to preserve their natural heritage. Over the last 40 years the agricultural policies of the European Union have changed dramatically—towards supporting more environment-friendly practices and less intensive use of soils and lands. But some aspects of its agricultural policy continue to work to destroy ecosystems.

The problem is clear, but the scale of the change required to address it is massive, and the efforts thus far token. Reforms will not be easy to formulate or apply and, most challengingly, we are yet to work out how to feed and support ourselves sustainably at the scale required. We all admire efficiency, but agricultural efficiency is starving many species out of existence. As noted by BfN (German Federal Agency for Nature Conservation) spokesman Franz August Emde: 'Farmers used to leave a few stalks standing. It gave field hamsters something to nibble on, and the birds benefited as well.'[6] In many places, farmers are once again being encouraged to leave corners of fields uncultivated or unharvested, and with financial assistance some large areas of farmland, such as the 1400-hectare Knepp Estate in the English Weald, are being retired from production and returned to nature.[7]

Nature has a remarkable capacity to recover from human-caused harm. Mark-Oliver Rodel, of Berlin's natural history museum, has been studying the reproductive behaviour of amphibians that live in places grossly disturbed by humans. They possess an astonishing ability to vary their reproductive patterns, and Rodel thinks that during 2000 years of disturbance, they have adapted genetically. I'm not surprised. After 90 million years, evolution must have taught them something about surviving in Europe.

Globalisation represents a different kind of threat to Europe's biodiversity. The Asian longhorned beetle, first detected in Italy in 2000, probably arrived in Europe in packing wood. It threatens a range of deciduous trees, including maple, birch and willow, none of which have adequate natural defences against it.[8] The larvae kill the trees by boring through the living wood; each one can consume up to a cubic metre of timber before it pupates.

The emerald ash borer is another Asian invader whose larvae destroy ash trees. These large and handsome beetles are just two of innumerable species of tree-infesting bacteria, fungi and invertebrates that have arrived in Europe in recent decades from that great evolutionary powerhouse Asia. As a result, almost every common type of European tree is now affected by one Asiatic disease or parasite or another. The process began 50 years ago when Europe's elms were ravaged by the beetle-borne fungal malady misnamed 'Dutch elm disease' (which, in fact, is Asian). More recently, sudden oak death, oak decline, beech wilt, sweet chestnut blight and horse chestnut canker have infested Europe's forests.

In the geological past, when the land bridge to temperate Asia was wide open, and climate favoured tree migration, the trees arrived alongside their pathogens. But because the bridge to Asia today is a human one, involving transported timber and plant seedlings and cuttings, the diseases are arriving in advance of the trees that can withstand them. According to the botanical writer Fiona Stafford, the only way to cope is to mimic what happened in times past, and plant Europe's forests with Asian varieties of the species at risk which, because they co-evolved with the diseases, are resistant to them.[9]

All of these changes are happening during the most rapid shift in climate in geological history. The current warming trend is at least 30 times faster than the warming that melted the great ice sheets at the end of the last glacial maximum, and the warming is increasing temperatures from one of the warmest points in the last three million years of Earth history. The cycle of ice ages has already been broken. Great ice sheets will not again advance across the north: the Pleistocene—one of the most tumultuous eras in Europe's turbulent geological history—is over.

Global temperatures are already 1° Celsius warmer than they were 200 years ago, and Europe is warming faster than the global average. Europe's forests and meadows are budding earlier in spring and birds are migrating earlier. Insects such as butterflies are not only appearing earlier but are reaching much further north than they used to. Climate change is a process, not a destination, and future changes will have far more impact. Both the Arctic tundra—a vital breeding ground for many species, including migratory geese—and alpine meadows are in danger of being smothered by forest. And that will mean farewell to the edelweiss, adieu the eider goose.

Even if the aspirations of the 2015 Paris Agreement on climate change are realised, Europe's coastline will alter, and some cities will be lost to rising seas. If the nations of the world fail to honour the pledges they made in Paris, the climate could return to Pliocene conditions, when okapi-like creatures and giant vipers thrived in Europe. It's a fair bet that Europe's agricultural productivity and political stability would then be imperilled. Ernst Haeckel's name for our Neanderthal ancestors, *Homo stupidus*, may yet have some validity—for us.

CHAPTER 43

Rewilding

What might Europe be? A new concept in the human management of natural systems is now emerging. Rewilding—restoring wild creatures and lost ecological processes—is becoming popular globally, but its origins are European, and it is there that the most substantial efforts at its realisation are being made. The independent organisation Rewilding Europe is carrying out extensive programs. Its objective is to restore natural process to ecosystems, to create wilderness areas where the human presence is minimal, and to introduce large herbivores and top predators to regions where they have been extirpated. The program intends to focus on ten areas, each one of at least 100,000 hectares, from western Iberia to Romania, Italy and Lapland.[*]

What do the Europeans imagine as they think of rewilding their continent? Some projects quote descriptions by the ancient Roman Tacitus, suggesting that once again Europeans are reaching into their dreamtime for inspiration. Others, however, think in geological time scales of tens of thousands or millions of years. It is to be expected that everyone's idea of what a wild Europe should look like will be a little different, but some agreement on baselines must be reached.

---

[*]The Rewilding Europe organisations are WWF-Netherlands, ARK Nature, Wild Wonders of Europe and Conservation Capital.

Should rewilders work to create the Europe of the Roman era, or 20,000, or (in light of climate change) two million years ago? Each would result in a wildly different outcome. Having set a baseline, rewilders could then work out which of the relevant species still exist, what their ecological requirements are, and what current species might be able to stand in as ecological substitutes for extinct kinds. They could also work out the minimum area required for such species, set about isolating and removing any species that should not be there, and reintroduce the species selected. Rewilding Europe, as it is currently being practised, is not quite that methodical. Some rewilders just let nature take its course with minimal intervention, while others focus largely on establishing three species of megafauna—the wisent, the wild horse, and the aurochs—which are, incidentally, the species most abundantly represented in European ice-age art, and which the ancient writers Tacitus, Herodotus and others saw in Europe, often in great herds.

Rewilding is not entirely new, nor is its history entirely honourable, for the first efforts were made by the Nazis in the 1930s and 40s. The brothers Lutz and Heinz Heck were German zoo directors, in Berlin and Munich respectively. Lutz joined the SS in 1933. He became a close friend of Hermann Göring and was obsessed with his own, perverted version of the great European dreamtime—a wilderness where Aryans could hunt dangerous wild animals of the sort he imagined the Teutonic tribes pursued in Roman and pre-Roman times.

An integral part of Lutz's program was the reconstitution of the aurochs, so that the master race could have its own special beast to hunt, one strong and dangerous, as befitting the ideal Aryan man. Starting with 'primitive breeds' of domestic cattle, Lutz and his brother Heinz selected not only for size and shape, but aggressiveness, which is why all but a few of Heck's cattle were later killed. They do not closely resemble the aurochs genetically. Attempts to re-create the aurochs have

started up again. The Tauros Program, supported by a Dutch foundation and a group of universities, is experimenting with eight ancient
European breeds using recently available DNA technologies, to identify
and selectively breed for animals with a high proportion of European
aurochs DNA. At the end of 2015 the project had yielded more than
300 crossbred animals, fifteen of which are fourth-generation crosses.
Ultimately, the program leaders hope to release 're-created' aurochs into
wild lands, where they can roam relatively freely.[1]

Lutz Heck decided that the Białowieża Forest was the perfect
location to create his great wilderness. The Nazis killed or chased
out thousands of people, destroying more than 300 villages. Among
their victims were the many Jews who had taken refuge in the dense
woods. Today Białowieża is prized as a World Heritage Site, testament
of the supposedly primitive, untouched plains forest of old Europe. We
forget the role the Nazis had in creating it, and the fact that the area
had previously been heavily inhabited and utilised for farms and forest
products for centuries. With the people gone, Heck released wisent,
bear and Heck cattle into the area, though it is doubtful that the Nazis
got much time to hunt. By May 1945 the Russians were in Berlin, and
Heck was busy defending his zoo, which acted as one of the last Nazi
redoubts in the city. At the end of the war the Russians tried to charge
Heck with war crimes, but he never faced the courts. He died on
6 April 1983 in Wiesbaden.

Lutz Heck's imagining that Europe was once covered with a great,
primeval forest was inspired in part by *Germania*, written about 98 CE.
In it the Roman historian Tacitus describes Germany as being covered
with *sylvum horridum*. Both Adolf Hitler and Heimlich Himmler,
incidentally, tried unsuccessfully to obtain the only surviving medieval
copy of the work—the *Codex Aesinas*—from its owners, the Counts of
Jesi in Ancona. But what did Tacitus mean by *sylvum horridum*? Was
Germany one great primeval forest, or was it covered in groves and

thickets of spiny plants, that could have been created by great herds of herbivores?[2]

Elsewhere, Tacitus leaves no doubt that parts of Germany were heavily modified landscapes, supporting crops, herds and villages. But he also says that each tribal area was surrounded by a sort of extensive no-man's-land. It's easy to imagine that these areas acted as hunting reserves, wherein wildlife was protected, to some extent, by one tribe's fear of being ambushed by another. Perhaps broken woodland interspersed with swamps and thorny thickets did characterise these areas. Europe's great diversity of light-requiring plants—the hazelnut, hawthorn and oak included—further supports the existence of an interrupted forest canopy in Europe. To mourn the loss of European sylvan virginity, as represented by those supposedly umbrageous Teutonic forests of Tacitus' *sylvum horridum*, is almost certainly a mistake.

The one good thing that seems to have come from Lutz Heck's obsessions was the survival of Warsaw Zoo's Przewalski's horses. He moved them to safety in his brother Heinz's zoo in Munich. By 1945 there were just 13 Przewalski's horses left on Earth, so Lutz's role was crucial to the species' survival—as has been recognised by at least one Holocaust museum.[3] Heck's attempted rewilding serves to reinforce a very important fact: Europeans are now the mind over their land. What they desire, the land will become. And if their desires are toxic and dangerous, then that will manifest itself in nature. Europeans cannot escape responsibility for shaping their environment; as even withdrawal from management will have profound consequences.

The idea that ancient Europe was a great primal forest is being challenged by one of its greatest rewilding projects—at Oostvaardersplassen in the Netherlands. In April 2017, I travelled there to meet Frans Vera, an ecologist who has had a huge hand in its development. The nearly 60 square kilometres that Frans and his colleagues have helped shape is less than 70 years old. Before that, it was under the sea. I found this extraordinary, but the Dutch are so used to creating their own land that it was barely commented on during my tour. Looking out on the vast tract through the misty morning air, with nothing but ghostly silhouettes of modern windmills and industrial buildings on the horizon, I felt that I had travelled back in time, the scene being reminiscent of untouched Africa, or the remote Arctic.

Being on the ground in the Oostvaardersplassen is a sensuous experience. The spoor and turds of birds and beasts lay so dense on the short-cropped sward that it was impossible to step between them. And the sward was so thin in early spring that there appeared to be more bare soil than grass. I could hardly believe that it supported such a mass of wildlife. As I watched, tens of thousands of barnacle geese lifted from it, as an enormous sea eagle soared overhead, then settled back like a mantle on the land once the danger passed. The smell, the sounds, and the sights might have been those of a Pleistocene European richness so long lost that it has vanished from our imaginations.

But the pride of the Oostvaardersplassen is its large mammals. Konik ponies cantered past us in small harems or family groups, their beautiful dun-coloured coats creating the illusion that I was looking at an animated panel of ice-age art. To an untrained eye, Koniks look almost like wild horses, and they were once thought to be descended from the tarpan—Europe's last wild horse.[4] Britain's Exmoor ponies, with their white muzzles—a feature so clearly seen in depictions of European ice-age horses in cave art—provide a further simulacrum. But this is really an aesthetic matter, for no living breed of horse is genetically closer to the wild ancestor than another.

A herd of red deer, led by a magnificently antlered stag, looked up as we passed, then sprang off. Their bones littered the ground. As a non-domesticated species, their carcasses are the only ones allowed to lie in the sward to feed scavengers. In the distance were great beasts, their heads bearing eerily familiar lyre-shaped horns. They are aurochs substitutes, bred from various races of domestic cattle. Some lack the uniform, dark coat of the aurochs, breaking, to my eye at least, the illusion of an ice-age megafauna.

The Oostvaardersplassen has exceptionally rich alkaline soil—it's the sort of place coveted by agriculturalists. There are no boulders behind which saplings might be sheltered, and the grasses of the rich soil support so many large mammals, which in turn determine what grows there. The result is a great sweetgrass sward, broken only in its lowest parts by waterlogged reedbeds. The things that are very obviously missing are trees. The few trees that are there are in terrible shape. Having been ring-barked by deer, their skeletal frames dot the land over vast areas, lending a funereal aspect. Beyond the odd blackthorn or hawthorn bush that a thousand pairs of lips had bonsaied to within an inch of its life, little living vegetable matter was taller than my ankle. Could that thorny bonsai, I wondered, be Tacitus's famed *sylvum horridum*?

In terms of its ecology, the Oostvaardersplassen, with its 4000-odd cattle, horses and deer, is reminiscent of the mammoth steppe or the short grass sward of the Masai Mara. Many see it as a failed experiment. Others simply hate the dead trees. I beg them to compare the Oostvaardersplassen not with their dreamtime Europe of the classical age, but with a long-vanished continent where large mammals, rather than agricultural practices, shaped landscapes.

In the creation of the Oostvaardersplassen some things have been lost, including 37 per cent of the bird species that existed there in 1989; most were adapted to an agricultural or partially forested Europe.[5] But in my view much more has been gained. The Oostvaardersplassen evokes a grand and wild Europe, a mini-version of the wildebeest migration on the Serengeti; but there is one big difference. The Oostvaardersplassen lacks large predators, foxes being the largest canid in the reserve. The exclusion of carnivores has had several implications, including what may be an unnatural density of herbivores. Another is that people have had to take the place of wolves and big cats. For humane reasons rangers stalk the area, especially in winter, and shoot animals judged to be too weak to survive until spring.

Nature continues to lead in the Oostvaardersplassen. An Egyptian vulture discovered the place and flourished until it was killed when it perched on a railway line. Will wolves or golden jackals also find their way in? Three wolves have been seen in the Netherlands already, so it seems possible. Even a moose—a zoo escapee—briefly made a home there. She had two young, but like the vulture she wandered onto the railway track and was killed, and then one of her young was shot. Perhaps wild boar will be the first large mammals to get to the Oostvaardersplassen under their own steam, for they are already at Nobelhorst, just a few kilometres away. If they do, they will discover a feast of bird eggs and other delicacies. And so the great experiment goes on. If it were up to me, I'd deal with that killer railway line first— either by inclosing it or rerouting it.

There were once plans to join this great plain up with other nature reserves in the Netherlands, and with wild areas in Germany, allowing natural migration, The Dutch government had acquired much of the land needed, but then a right-wing government was elected. Farmers cried that the rich soils were being wasted, and some were allowed to buy back land they had sold, at lower prices than they were paid for it. The political negativity confused the public, and so a grand vision was destroyed. I hope that the vast experiment that is the Oostvaardersplassen continues. We learn more every year as it provokes imaginative responses that help guide the mind over the land in the most innovative ways.

At the other end of Europe, in Romania, a very different kind of experiment in rewilding is taking place. At its heart lies the

Carpathians, which form a curved mountainous and well-forested spine that provides habitat for one-third of Europe's bears, along with many other wild species. In Romania, even in farmed areas, wildlife abounds, and magnificent wildflower meadows bloom in spring. This is in part because older, less destructive agricultural practices persist, with shepherds still tending flocks, and horse power remaining common on farms and roads. Because of Romania's abundant carnivores, roe deer, rabbits and red deer are hard to find.

The Conservation Carpathia Foundation is a not-for-profit organisation that has a small holding of about 400 hectares of pastureland near the village of Cobor in Transylvania. I stayed there in April 2017 to learn how the organisation acts as a model for environmental farming in the region. Christoph Promberger, the executive director, told me about the organisation's really big project, which is located in the Făgăraş Mountains, arguably the wildest region in Europe.

The Făgăraş region is extraordinarily rugged and beautiful, combining a Swiss-like landscape with substantial populations of bears, wolves, lynx, roe deer and red deer. With the nearest village 40 or more kilometres away from the project site, the forests are as remote as anywhere can be in Europe. Yet they came under severe threat after Nicolae Ceauşescu was deposed. Romania's forests had been nationalised, but in the early years of the post-communist era, the previous owners were each given back a hectare of their estate. A few years later that was expanded to 10 hectares, and in 2005 their entire holdings were restored. Uncertain whether their lands would be taken again, most owners proceeded to clear-cut the trees to make a quick profit. In order to avoid a total catastrophe, Conservation Carpathia began to buy up the reprivatised forest lands.

Already, Conservation Carpathia owns 15,000 hectares of forest or recently logged land, and it has plans to purchase 45,000 hectares more. There are proposals to create a national park of nearly 200,000 hectares in the Făgăraș. If this were to be realised, combined with the lands purchased by Conservation Carpathia, the area would become the largest wilderness in Europe. Rewilding Europe has already released wisent into the Carpathians, and Conservation Carpathia also plans, in 2018, to re-introduce wisent. Because there are no plans to reintroduce other species, the Făgăraș ecosystem will lack some vultures and eagles, wild horses and aurochs (or their equivalent), not to mention the great beasts of ice-age Europe. But like Oostvaardersplassen, it promises to be a most interesting experiment.

The Oostvaardersplassen and Făgăraș are bookends in a great, pan-European project to rediscover the continent's nature. Both are worth refining and amplifying. We should be in no rush to rewild Europe; neither should we ignore some important challenges, one of the greatest being the scavenging role. Despite the hyena's long history in Europe, nobody seems to want to return it to the continent, and vultures are extinct over much of the landmass, with attempts to reintroduce them falling foul of many obstacles, from bureaucracy to powerlines, railway tracks, poisoned baits and pesticides. The only vulture seen in Romania in recent years died after drinking water contaminated with pesticides.

Not all rewilding results from sanctioned actions. In 2006 a small population of beavers mysteriously appeared on the River Otter, in Devon. Someone must have released them without permits or public discussion. The authorities wanted them removed, but the locals liked having beavers around and kicked up a fuss, so the proposed eradication was dropped. The British have a reputation for resenting rules, so perhaps we should anticipate further unplanned introductions. But surely the potential also exists in eastern Europe, in places like Russia, where regulation is looser, and where great wealth resides in the hands of a few.

CHAPTER 44

Re-creating Giants

Much of Europe's megafauna has, like the fabled trolls and goblins of mythology, retreated long ago to distant or invisible realms: so it is that relatives of Europe's extinct elephants roam unrecognised in the forests of the Congo, while the genes of aurochs, cave bears and Neanderthals lie hidden in the genomes of cattle, brown bears and humans. And in the far north, the DNA of woolly mammoth and woolly rhinos lie in perpetual sleep, rocked in the bosom of the permafrost. Clever goblins, working in idea-factories, have stumbled across the magic required to return these vanished giants to their ancestral homes—whether by introduction, selective breeding or genetic manipulation. If the Europeans think small, Europe will remain a diminished place—one shorn of its greatest natural glories. But, if they think big, anything is possible.

Europe's vanished wildlife falls into four categories: 1) those that survive as living creatures outside Europe; 2) those that can be re-created through the selective breeding of domestic stock; 3) those that it might be possible to re-constitute through genetic engineering; 4) those that, given current technology and knowledge, are irrecoverable.

The easiest species to restore are those that survive elsewhere: spotted hyena, lion, leopard, water buffalo and arguably straight-tusked elephant (also known as African forest elephant), to name a few, are all vanished from Europe, but can be found in Africa or Asia. The next easiest to restore are those that can be resurrected through selective breeding, but only the aurochs, the European wild horse and the Neanderthal fall into this category. From a technical perspective, de-extincting a Neanderthal would be the easiest task of all, for human reproduction is extremely well understood, and the Neanderthal genome is known. But the last people to try selective breeding involving humans were the Nazis, and the idea is utterly immoral—I'm sure the ghost of Lutz Heck would look on with great interest.

Among the irrecoverable species must be counted Europe's three rhinos (the woolly, Merck's and the narrow-nosed), the giant elk, and island species such as *Myotragus*. But the study of ancient DNA is developing rapidly, and before too long the genomes of several species may well be recovered. The third category—the species that may be recovered through genetic engineering—takes us to the outer limits of scientific knowledge. In 2008 there was an attempt to revive the Spanish bucardo, a subspecies of ibex. The last individual had died in 2000, but scientists had taken ear clippings from her the previous year. They transplanted DNA from the frozen clippings into the cells of domestic goats. One of the embryos thus created survived to birth, but the young bucardo died just seven minutes later, after breathing difficulties.[1]

Among extinct species the principal candidates for genetic restoration are, in terms of feasibility, the woolly mammoth, cave bear and cave lion. Revive and Restore is an organisation dedicated to using genetics to save endangered species, and to restore species from extinction.[2] It is working on a diverse range of projects, from assisting with the increased use of a synthetic substitute for horseshoe-crab blood (the creatures are overharvested for their blood, which is used in the pharmaceutical industry), to supporting the Harvard Woolly Mammoth Revival Team.

In early February 2017 it was widely reported in global media that the woolly mammoth would be 'back from extinction' by 2018. In fact, George Church, who leads the Harvard Woolly Mammoth Revival Team, had claimed that by 2018, his team hoped to create a viable embryo—perhaps just a few cells—of a creature containing a composite of Asian elephant and woolly mammoth genes. A mammophant, if you like. Given what we now know about elephant hybridisation, this doesn't sound as outrageous as it might once have. Indeed, perhaps we should see CRISPR technology (a technology enabling genes from one species to be inserted into another) as a continuation of elephant evolution through hybridisation, as has been the case for millions of years.

Even this more limited ambition, however, speaks eloquently of the rapid progress being made in the area of de-extinction. Church and his team plan to create the mammophant by endowing the egg cell of an Asian elephant with the genes for red blood cells that operate efficiently at low temperature, an enhanced fat layer under the skin, and a luxuriant covering of hair and fur—all from the woolly mammoth genome. The team has already made 45 changes out of the 1642 differences between the elephant and mammoth genomes. But this is just the beginning. The nuclear DNA must then be placed in an embryo, much as Dolly the sheep's nuclear DNA was replaced to create the world's first cloned sheep.

The team does not intend to use an elephant egg cell taken from an elephant, but rather to create one from skin cells. Finally, the growing embryo would need to be kept in an artificial womb for twenty-two months, before a baby mammophant could be produced. And from there a genetically mixed, appropriately age-structured herd of mammophants would need to be 'manufactured' if the 'species' is to be restored to its ecosystem.[3] I have little doubt that, given time, all of this will be possible. But first, humanity must decide whether it's desirable.

The genetic reincarnation of Europe's lost giants would not be the last step, for a sufficiently large and fertile area would have to be set aside for hundreds if not thousands of mega-mammals to roam. Europe is hardly ideal for re-creating the mammoth steppe. But a large project in Siberia, aimed at doing this, is already well underway.* If the mammoth can be restored, then so in all probability can the cave bear and the cave lion. But what would be gained by re-creating them? If a top predator is someday required for a European rewilding project, the living lion is probably a better candidate than the cave lion, being adapted to the prevailing warmer conditions. And by exerting selective pressure on the brown bear to make it herbivorous, we have effectively re-created a large herbivorous ursine that probably occupies the cave bear's ecological niche. If Europe is to rewild in this age of warming, it is the temperate species, such as the lion and the straight-tusked elephant that must be the focus, and even the largest temperate wilderness on the continent is currently too small for them. But by 2030 it is predicted that there will be 30 million hectares of abandoned farmland in Europe.[4] Most of Europe's national parks exist on privately owned land, and European landowners tend to accept the imposition of societal decisions. It is to the flexible and adaptive European concept of land ownership and the opportunities opened by land abandonment that future generations must turn if they wish to realise the nascent dream of a dynamic, megafaunal Europe.

But should the Europeans seek to re-create a European megafauna by importing species similar to those it once had, which survive elsewhere? I think the moral case is unassailable: it is unacceptable to ask the people of Africa, whose population may reach around four billion by 2100, to live alongside lions and elephants while Europeans refuse

---

* The project, known as Pleistocene Park, is being led by Sergei Zimov. Horses, moose, muskox, wapiti and wisent are already present in the enclosure.

to do so. If we ask others to shoulder such a disproportionate burden, I fear that there will no longer be a place for elephants in this world.[5]

The scale of land abandonment in Europe is already so great that managed rewilding is being conducted on only a minute fraction of abandoned lands. Most is instead subject to a huge unplanned experiment, with little or no scientific oversight, and in which an array of species there by happenstance are shaping the future. In the mountainous Colline Metallifere of Grosseto and Siena provinces in Italy, for example, land abandonment is creating a vast new wilderness, which is currently of exceptional diversity. Despite its location amid the manicured landscapes of Tuscany, the region has the lowest population density in all of Italy, and some of its greatest biodiversity. Maquis grows on the warmer slopes, and elsewhere a very diverse forest including oak, holly, chestnut and aspen thrive. The understorey of the regrowing forest is cropped by roe, fallow and red deer—the last two species having escaped from captivity in recent decades. There are no lynx in Tuscany, so the deer population is dense and is having a severe impact on the understorey. Just a few unpalatable species, like juniper, now survive the seedling stage, and if nothing is done they will create a future, impoverished forest of the Colline Metallifere.

Some people believe that humans should not seek to guide the development of the ecosystems of Europe's newly wild lands, imagining that they will return to some primeval and desirable state if left alone. But it is already clear that this will not occur, and that a less-diverse and unproductive forest will result from the current haphazard mix of landscape architects comprising the large herbivores and carnivores. The big decisions, as far as human management goes, involve deciding what kinds of large herbivores and carnivores should be released into the unmanaged lands. To make wise decisions, a long-term view is needed.

Luigi Boitani lives amid the regrowing forests of Tuscany's Colline Metallifere. Soon after he moved there he planted an acorn beside his

house. Today it is a thriving sapling five metres high. I can imagine
the grand old tree that, given a little luck, it will become by 2030, but
both Luigi and I struggle to envision the forest it will exist in, let alone
the Europe of 180 years hence. The only thing that we're sure of is that
there will be plenty of surprises.

Let's enter our time machine for one last journey—into an imagined future Europe 180 years from now, to visit Luigi's oak in its
maturity. We approach a continent that in one respect looks like the
archipelago of old: cities stand out like islands linked with transport
corridors, each surrounded by a penumbra of greenhouses and other
enclosed structures that produce the food the population requires.
Instead of being separated by sea, Europe's cities are separated by
vast areas of forests and woodlands—the result of centuries of land
abandonment. We touch down beside Luigi's oak, which is growing
in a grassy woodland surrounded by palms, ginkgoes and magnolias,
as well as chestnuts, oaks and beech; courtesy of climate change, the
Arcto-Tertiary Geoflora is well on its way to re-establishing in Europe.

Before us in the glade stand two statues. One honours a twenty-
first century Russian oligarch who released his immense collection of
wild animals into abandoned lands in eastern Europe. Thanks to him,
Europe once again has lions, spotted hyena and leopards. A second
statue honours a long-sighted Dutch woman who crowdfunded a
project to gather the world's last Sumatran rhinos and straight-tusked
elephants and release them in a fenced estate created from recently
vacated farmland in western Europe. Given food and shelter, they
adapted to the new climate. Eventually the fence was taken down, and
elephants and rhinos once more roamed European forests.

A group of tourists from Africa and Asia, hoping to see elephant
and rhino, is led by a young European tour guide. She explains that
once-upon-a-time Africa and Asia had megafauna too, but they did not
survive the booming population and political chaos in the twenty-first

century. She points out an elephant with mammoth-like features. It's a mammophant whose mixed genetic heritage allows it to fill the ecological niche of a mammoth and still survive in Europe's warmer climate. The guide explains that scientists discovered that Europe's ecosystems required two elephant species if they were to stay diverse and healthy, so the mammophant was genetically engineered. The first specimens learned behaviours necessary for their survival after being adopted into herds of straight-tusked elephants. But now there are enough of them to form herds of their own.

The guide is armed only with a small, high-technology stick, yet she is completely at ease with the great beasts around her—much like Australian tour guides in the land of sharks and crocodiles. And it is this ease with nature that the Europeans have become famous for. As many young Europeans live in the complex ecosystems they have helped to create as do in the cities, for the forests offer adventure, and the possibility of learning something new. The lifestyles of the Europeans are very different from those of the rest of the world's population, which is concentrated in mega-cities without access to wild areas. A dynamic and adventurous people, the Europeans are always thinking of something new.

Envoi

In the German city of Worms, a medieval carving depicts a woman holding a midwife toad, which denotes that the woman is herself a midwife.[1] The Europeans are the eternal midwives of their environment: every interaction they have with it helps give birth to a new Europe. Let us hope that this generation are midwives with vision.

Acknowledgments

Luigi Boitani contributed much material concerning Europe over the past millennium and brought his unparalleled knowledge of European carnivores and the dilemmas of managing abandoned land in Europe to the project. We do not agree on every sentiment expressed in the book. Any errors are mine, and contentious points of view my responsibility.

Kate and Coleby Holden accompanied me on the many journeys required to write this book. Kate read the manuscript and provided many useful comments. I owe Brian Rosen an enormous debt of gratitude for sharing his profound knowledge of European geology and palaeontology. Kris Helgen read the entire manuscript and corrected many errors. Jerry Hooker generously shared his research into early mammals, and was exceedingly generous with his time, illuminating many aspects of European prehistory and palaeontology. Colin Groves critiqued the first third of the manuscript in the last week of his life, in his usual acute and humorous manner, and Martin Aberhan and Johannes Müller, both herpetologists, explained their important research. Some of the writing and research for this book was completed while I taught at the Graduate Institute in Geneva. Its director, Professor Philippe Burrin, provided much stimulating conversation and encouragement. A special thanks to Claudio Segre, for supporting me while at the Graduate Institute, and for his wonderful hospitality. In Romania, Enrico Perinyi and the staff of Seneca Publishing, especially Anastasia, Irina, Catiline, Micale, Maria and Christie, made our visit a most enjoyable and enlightening experience. The staff of Wildlife Carpathia and the Hateg Geopark were also extremely generous with their time. Dr Valentin Paraschiv, Dr Dan Grigorescu and Dr Ben Kear all deserve my thanks for assisting with information and discussions.

Nick Rowley alerted me to the plight of Europe's smaller birds, and Geoff Holden informed me of many other matters, as well as reading a draft of the manuscript. Finally, thanks must go to my editors, Michael Heyward and Jane Pearson, at Text Publishing, who have made this a far better book.

Endnotes

INTRODUCTION

1   Wodehouse, P. G., *The Code of the Woosters*, Herbert Jenkins, London, 1938.

CHAPTER 1

1   Much of the rest of this chapter has been distilled from a recent and detailed review: 'Island Life in the Cretaceous—Faunal Composition, Biogeography, Evolution, and Extinction of Land-living Vertebrates on the Late Cretaceous European Archipelago', Zoltan Csiki-Sava, Eric Buffetaut, Attila Ősi, Xabier Pereda-Suberbiola, Stephen L. Brusatte, *ZooKeys* 469: 1–161 (08 Jan 2015). I am extremely grateful for their work in bringing so many scattered references together and placing them in context.

2   Signor III, P. W. and Lipps, J. H., 'Sampling Bias, Gradual Extinction Patterns, and Catastrophes in the Fossil Record', in Silver, L. T and Schultz, P. H. eds., *Geological Implications of Impacts of Large Asteroids and Comets on the Earth*, Geological Society of America Special Publications, Vol. 190, pp. 291–96, 1982. A taxon, incidentally, is a taxonomic grouping of organisms.

3   This reconstruction of Hateg's flora is drawn from a number of sources that document the flora of Modac and Bal over a long period of time. It thus paints a broadbrush picture, some details of which may not be precisely applicable to Hateg at the time that some of the creatures discussed existed.

4   Blondel, J. *et al*, *The Mediterranean Region: Biological Diversity in Space and Time*, Oxford University Press, Oxford, 2010, 2nd edition, Chapter 3.

CHAPTER 2

1   Veselka, V., 'History Forgot this Rogue Aristocrat Who Discovered Dinosaurs and Died Penniless', *Smithsonian Magazine*, July 2016,

http://www.smithsonianmag.com/history/history-forgot-rogue-aristocrat-discovered-dinosaurs-died-penniless-180959504/

2   Gaffney, E. S. & Meylan, P. A., 'The Transylvanian Turtle Kallokibotion, a Primitive Cryptodire of the Cretaceous Age', *American Museum Novitates*, 3040, 1992.

3   *Ibid.*

4   Edinger, T., 'Personalities in Palaeontology—Nopcsa', *Society of Vertebrate Palaeontology News Bulletin*, Vol. 43, pp. 35–39, New York, 1955.

5   *Ibid.*

6   Taschwer, K., 'Othenio Abel, Kämfer gegen die "Verjudung" der Universität', *Der Standard*, 9 October 2012.

7   *Ibid.*

8   Nopcsa, F., 'Die Lebensbedingungen der Obercretacischen Dinosaurier Siebenbürgens', *Centralblatt für Mineralogie und Paläontologie*, Vol. 18, pp. 564–574, 1914.

9   Plot, R., *The Natural History of Oxfordshire, Being an Essay towards the Natural History of England*, Printed at The Theatre in Oxford, 1677, llustration on p. 142, discussion, pp. 132–36.

10 Brookes, R., *A New and Accurate System of Natural History: The Natural History of Waters, Earths, Stones, Fossils, and Minerals with their Virtues, Properties and Medicinal Uses, to which Is Added, the Method in which Linnaeus has Treated these Subjects,* J. Newberry, London, 1763.

11 International Commission on Zoological Nomenclature, http://iczn.org/iczn/index.jsp

12 Edinger, T., 'Personalities in Palaeontology—Nopcsa', *Society of Vertebrate Palaeontology News Bulletin*, Vol. 43, pp. 35–39, New York, 1955.

13 Colbert, E. H., *Men and Dinosaurs*, E. P. Dutton, New York, 1968.

14 Veselka, V., 'History Forgot this Rogue Aristocrat Who Discovered Dinosaurs and Died Penniless', *Smithsonian Magazine*, July 2016.

CHAPTER 3

1   Nopcsa, F., 'Die Dinosaurier der Siebenbürgischen Landesteile Ungarns', *Mitteilungen aus dem Jahrbuch der Ungarischen Geologischen Reichsanstalt*, Vol. 23, pp. 1–24, 1915. Unsurprisingly, Abel dismissed this work.

2   Colin Groves, personal communication. The skeleton was in fact a composite that had been made up of the bones of several individuals.

3   Thomson, K.,'Jefferson, Buffon and the Moose', *American Scientist*, Vol. 6, No. 3, pp. 200–02, 2008.

4   Buffetaut, E. *et al,* 'Giant Azhdarchid Pterosaurs from the Terminal Cretaceous of Transylvania (Western Romania)', *Naturwissenschaften*, Vol. 89, pp. 180–184, 2002.

5   Panciroli, E, 'Great Winged Transylvanian Predators Could have Eaten Dinosaurs', *Guardian*, 8 February 2017.

CHAPTER 4

1   Skelton, T. W., *The Cretaceous World*, Chapter 5, Cambridge University Press, 2003.

2   Koch, C. F. and Hansen, T. A., 'Cretaceous Period Geochronology', *Encyclopaedia Britannica*, 1999.

CHAPTER 5

1   Darwin, C., *On the Origin of Species by Means of Natural Selection, or the Preservation of Favoured Races in the Struggle for Life*, John Murray, London, 1859.

2   Zhang, P. *et al*, 'Phylogeny and Biogeography of the Family Salamandridae (Amphibia: Caudata) Inferred from Complete Mitochondrial Genomes', *Molecular Phylogenetics and Evolution*, Vol. 49, pp. 586–97, 2008.

3   *Ibid.*

CHAPTER 6

1   Mayol, J. *et al*, 'Supervivencia de Baleaphryne (Amphibia: Anura: Discoglossidae) a Les Muntanyes de Mallorca', nota preliminar, Butll. Inst. Cat, Hist. Nat., 45 (Sec. Zool., 3) pp. 115–19, 1980.

2   Koestler, A., *The Case of the Midwife Toad*, Random House, New York, 1971.

3   Semon, R., *Die mnemischen Empfindungen*, William Engelmann, Leipzig, 1904; English translation: Semon, R., *The Mneme*, George Allen & Unwin, London, 1921. Both Sigmund Freud and the Church of Scientology borrowed heavily from Semon's ideas.

4   Cock, A. and Forsdyke, D. R., *Treasure Your Exceptions: The Science and Life of William Bateson*, Springer-Verlag, New York, 2008.

5   Raje, J.-C. and Rocek, Z., 'Evolution of Anuran Assemblages in the Tertiary and Quaternary of Europe, in the Context of Palaeoclimate and Palaeogeography', *Amphibia-Reptilia*, Vol. 23, No. 2, pp. 133–67, 2003.

CHAPTER 7

1   Vila, B. *et al*, 'The Latest Succession of Dinosaur Tracksites in Europe: Hadrosaur Ichnology, Track Production and Palaeoenvironments', *PLOS ONE*, 3 September 2013.

2   Perlman, D., 'Dinosaur Extinction Battle Flares', *Science*, 7 February 2013.

3   Keller, G., 'Impacts, Volcanism and Mass Extinction: Random Coincidence or Cause and Effect', *Australian Journal of Earth Sciences*, Vol. 52, pp. 725–57, 2005.

4   Sandford, J. C. *et al*, 'The Cretaceous–Paleogene Boundary Deposit in the Gulf of Mexico: Large-scale Oceanic Basin Response to the Chicxulub Impact', *Journal of Geophysical Research*, Vol. 121, pp. 1240–61, 2016.

5   Yuhas, A., 'Earth Woefully Unprepared for Surprise Comet or Asteroid, Nasa Scientist Warns', *Guardian*, 13 December 2016.

CHAPTER 8

1   International Commission on Stratigraphy, International Union of Geological Sciences, www.stratigraphy.org/index.php/ics-chart-timescale

2   Labandeira, C. C. *et al*, 'Preliminary Assessment of Insect Herbivory across the Cretaceous–Tertiary Boundary: Major Extinction and Minimum Rebound', in Hartman, J. H. *et al*, eds., *The Hell Creek Formation and the Cretaceous–Tertiary Boundary in the Northern Great Plains: An Integrated Continental Record of the End of the Cretaceous*, Geological Society of America, 2002.

3   De Bast, E. *et al*, 'Diversity of the Adapisoriculid Mammals from the Early Paleocene of Hainin, Belgium', *Acta Palaeontologica Polonica*, Vol. 57, No. 1, pp. 35–52, Warsaw, 2012.

4   Taverne, L. *et al*, 'On the presence of the Osteoglossid Fish Genus *Scleropages* (Teleostei, Osteoglossiformes) in the Continental Paleocene of

Hainin (Mons Basin, Belgium)', *Belgian Journal of Zoology*, Vol. 137, No. 1, pp. 89–97, Royal Belgian Institute of Natural Sciences, Brussels, 2007.

5   Delfino, M. and Sala, B., 'Late Pliocene Albanerpetontidae (Lissamphibia) from Italy', *Journal of Vertebrate Paleontology*, Vol. 27, No. 3, pp. 716–19, Society of Vertebrate Paleontology, New York, 2007.

6   Puértolas, E. *et al*, 'Review of the Late Cretaceous–Early Paleogene Crocodylomorphs of Europe: Extinction Patterns across the K–PG Boundary', *Cretaceous Research*, Vol. 57, pp. 565–90, 2016.

7   Folie, A. & Smith, T., 'The Oldest Blind Snake Is in the Early Paleocene of Europe', Annual Meeting of the European Association of Vertebrate Palaeontologists, Turin, Italy, June 2014.

8   Folie, A. *et al*, 'New Amphisbaenian Lizards from the Early Paleogene of Europe and Their Implications for the Early Evolution of Modern Amphisbaenians', *Geologica Belgica*, Vol. 16, No. 4, pp. 227–35, 2013.

9   Longrich, N. R. *et al*, 'Biogeography of Worm Lizards (Amphisbaenia) Driven by End-Cretaceous Mass Extinction', Proceedings of the Royal Society B, Vol. 282, Issue 1806, 2015.

10 Kielan-Jaworowska, Z. *et al*, *Mammals from the Age of Dinosaurs: Origins, Evolution, and Structure*, Columbia University Press, New York, 2004.

11 Smith, T. and Codrea, V., 'Red Iron-Pigmented Tooth Enamel in a Multituberculate Mammal from the Late Cretaceous Transylvanian "Hateg Island"', *PLOS ONE*, Vol. 10, No. 7, San Francisco, 2015.

12 De Bast, H. *et al*, 'Diversity of the Adapisoriculid Mammals from the Early Paleocene of Belgium', *Acta Palaeontologica Polonica*, Vol. 57, pp. 35–52, Warsaw, 2011.

CHAPTER 9

1   Malthe-Sørenssen, A. *et al*, 'Release of Methane from a Volcanic Basin as a Mechanism for Initial Eocene Global Warming', *Nature*, Vol. 429, pp. 542–45, 2004.

2   Cui, Y. *et al*, 'Slow Release of Fossil Carbon during the Paleocene–Eocene Thermal Maximum', *Nature Geoscience*, Vol. 4, pp. 481–85, 2011.

3   Beccari, O., *Wanderings in the Great Forests of Borneo*. A Constable & Co, London, 1904.

4   Hooker, J. J., 'Skeletal Adaptations and Phylogeny of the Oldest Mole *Eotalpa* (Talpidae, Lipotyphla, Mammalia) from the UK Eocene: The

Beginning of Fossoriality in Moles', *Palaeontology*, Vol. 59, Issue 2, pp. 195–216, 2016.

5 He, K. *et al*, 'Talpid Mole Phylogeny Unites Shrew Moles and Illuminates Overlooked Cryptic Species Diversity', *Mol. Biol. Evol.* Vol. 34, Issue 1, pp. 78–87, 2016.

6 Hooker, J. J., A Two-Phase Mammalian Dispersal Event Across the Paleocene–Eocene Transition', *Newsletters on Stratigraphy*, Vol. 48, pp. 201–20, 2015. (The elephant shrew genus in question is *Cingulodon*.)

7 De Bast, E. and Smith, T., 'The Oldest Cenozoic Mammal Fauna of Europe: Implications of the Hainin Reference Fauna for Mammalian Evolution and Dispersals during the Paleocene', *Journal of Systematic Palaeontology*, Vol. 19, No. 9, pp. 741–85, Natural History Museum, London, 2017.

8 Mayr, G., 'The Paleogene Fossil Record of Birds in Europe', *Biological Reviews*, Vol. 80, Issue 4, pp. 515–42, Cambridge Philosophical Society, 2005.

9 Angst, D. *et al*, 'Isotopic and Anatomical Evidence of an Herbivorous Diet in the Early Tertiary Giant Bird Gastornis: Implications for the Structure of Paleocene Terrestrial Ecosystems', *Naturwissenschaften*, Vol. 101, Issue 4, pp. 313–22, Springer-Verlag, New York, 2014.

10 Folie, A. *et al*, 'A New Scincomorph Lizard from the Palaeocene of Belgium and the Origin of Scincoidea in Europe', *Naturwissenschaften*, Vol. 92, Issue 11, pp. 542–46, Springer-Verlag, New York, 2005.

11 *Ibid*.

12 Russell, D. E. *et al*, 'New Sparnacian Vertebrates from the "Conglomerat de Meudon" at Meudon, France', *Comptes Rendus*, Vol. 307, pp. 429–33, Académie des Sciences, Paris, 1988.

CHAPTER 10

1 Switek, B. 'A Discovery that Will Change Everything (!!!) … Or Not', ScienceBlogs, 18 May 2009.

2 Strong, S. and Schapiro, R., 'Missing Link Found? Scientists Unveil Fossil of 47-Million-Year-Old Primate, *Darwinius Masillae*', *Daily News,* 19 May 2009.

3 Leake, J. and Harlow, J., 'Origin of the Specious', *Times Online*, 24 May 2009.

4 Amundsen, T. *et al*, 'Ida' er oversolgt, *Aftenposten* – Ida er en oversolgt bløff, *Nettavisen*, *Dagbladet*, 20 May 2009.

5   Cline, E. 'Ida-lized! The Branding of a Fossil', *Seed Magazine*, USA, 22 May 2009.

6   Hooker, J. J. *et al*, 'Eocene–Oligocene Mammalian Faunal Turnover in the Hampshire Basin, UK: Calibration to the Global Time Scale and the Major Cooling Event', *Journal of the Geological Society*, Vol. 161, pp. 161–72, March 2004.

7   Mayr, G., 'The Paleogene Fossil Record of Birds in Europe', *Biological Reviews*, Vol. 80, pp. 515–42, 2005.

8   Mayr, G., 'The Paleogene Fossil Record of Birds in Europe', *Biological Reviews*, Vol. 80, No. 4, pp. 515–42.

CHAPTER 11

1   Wallace, C. C., 'New Species and Records from the Eocene of England and France Support Early Diversification of the Coral Genus *Acropora*', *Journal of Paleonology*, Vol. 82, No. 2, pp. 313–28, 2008.

2   Duncan, P. M., *A Monograph of the British Fossil Corals*, Second Series, Part 1, 'Introduction: Corals from the Tertiary Formations', Palaeontographical Society, London, 1866.

3   *Ibid*.

4   Tang, C. M., 'Monte Bolca: An Eocene Fishbowl', in Bottiger, D. *et al*, (eds.), *Exceptional Fossil Preservation*, Columbia University Press, New York, 2002.

5   *Ibid*.

6   Bellwood, D. R., 'The Eocene Fishes of Monte Bolca: The Earliest Coral Reef Fish Assemblage', *Coral Reefs*, Vol. 15, pp. 11–19, 1996.

CHAPTER 12

1   Huyghe, D. *et al*, 'Middle Lutetian Climate in the Paris Basin: Implications of a Marine Hotspot of Palaeobiodiversity', Facies, Springer Verlag, Vol. 58, No. 4, pp. 587–604, 2012.

2   Gee, H., 'Giant Microbes that Lived for a Century', *Nature*, 19 August 1999.

3   Kirkpatrick, R., *The Nummulosphere: An Account of the Organic Origin of so-called Igneous Rocks and of Abyssal Red Clays*, Lamley and Co., London, 1913.

4   Waddell, L. M. and Moore T. C., 'Salinity of the Eocene Arctic Ocean from Oxygen Isotope Analysis of Fish Bone Carbonate', *Paleoceanography and Paleoclimatology*, Vol. 23, Issue 1, March 2008.

5   *Ibid.*

6   Barke, J. *et al,* (2012). 'Coeval Eocene Blooms of the Freshwater Fern Azolla in and around Arctic and Nordic Seas', *Palaeogeography, Palaeoclimatology, Palaeoecology*, Vol. 337–38, pp. 108–19, 2012.

CHAPTER 13

1   Sheldon, N. D., 'Coupling of Marine and Continental Oxygen Isotope Records During the Eocene–Oligocene Transition', *GSA Bulletin*, Vol. 128, pp. 502–10, 2015.

2   Hooker, J. J. *et al*, 'Eocene–Oligocene Mammalian Faunal Turnover in the Hampshire Basin, UK: Calibration to the Global Time Scale and the Major Cooling Event', *Journal of the Geological Society*, Vol. 161, pp. 161–72, March 2004.

3   Arkgün, F. *et al*, 'Oligocene Vegetation and Climate Characteristics in North-West Turkey: Data from the South-Western Part of the Thrace Basin', *Turkish Journal of Earth Sciences*, Vol. 22, pp. 277–303, 2013.

4   *Ibid.*

5   Mazzoli, S. and Helman, M. 'Neogene Patterns of Relative Plate Motion for Africa-Europe: Some Implications for Recent Central Mediterranean Tectonics', *Geol Rundsch*, Vol. 83, pp. 464–68, 1994.

6   Sundell, K. A., 'Taphonomy of a Multiple *Poebrotherium* Kill Site—an *Archaeotherium* Meat Cache', *Journal of Vertebrate Palaeontology*, Vol. 19, Supp. 3, 79a, 1999.

7   Pickford, M. and Morales, J., 'On the Tayassuid Affinities of *Xenohyus* Ginsburg, 1980, and the Description of New Fossils from Spain', *Estudios Geologicos*. Vol. 45, pp. 3–4, 1989.

8   Weiler, U. *et al*, 'Penile Injuries in Wild and Domestic Pigs', *Animals*, Vol. 6, No. 4, p. 25, 2016.

9   www.news.com.au/technology/science/animals/woman-mauled-by-viciousherd-of-javelinas-in-arizona/news-story

10   Menecart, B., 'The Ruminantia (Mammalia, Certiodactyla) of the Oligocene to the Early Miocene of Western Europe: Systematics, Palaeoecology and Palaeobiogeography', PhD thesis 1756, University of Fribourg, 2012.

CHAPTER 14

1   *Ibid*.

2   Mayr, G., 'The Paleogene Fossil Record of Birds in Europe', *Biological Reviews*, Vol. 80, pp. 515–42, 2005.

3   Mayr, G. and Manegold, A., 'The Oldest European Fossil Songbird from the Early Oligocene of Germany', *Naturwissenschaften*, Vol. 91, pp. 173–77, 2004.

4   Low, I., *Where Song Began: Australia's Birds and How They Changed the World*, Penguin Books Australia, Melbourne, 2014.

5   *Ibid*.

6   Naish, D., 'The Amazing World of Salamander', *Scientific American* blog, 1 October 2013.

7   Naish, D., 'When Salamanders Invaded the Dinaric Karst: Convergence, History and the Re-emergence of the Troglobitic Olm', *Tetrapod Zoology*, 17 November 2008.

1   Antoine, P. O. and Becker, D., 'A Brief Review of Agenian Rhinocerotids in Western Europe', *Swiss Journal of Geoscience*, Vol. 106, Issue 2, pp. 135–46, 2013.

CHAPTER 15

2   Campani, M. *et al*, 'Miocene Palaeotopography of the Central Alps', *Earth and Planetary Science Letters*, Vols. 337–38, pp. 174–85, 2012.

3   Jiminez-Moreno, G. and Suc, J. P., 'Middle Miocene Latitudinal Climatic Gradient in Western Europe: Evidence from Pollen Records', *Palaeogeography, Palaeoecology, Palaeobiology*, Vol. 253, pp. 224–41, 2007.

4   Čerňanský, A. *et al*, 'Fossil Lizard from Central Europe Resolves the Origin of Large Body Size and Herbivory of Giant Canary Island Lacertids', *Zoological Journal of the Zoological Society*, Vol. 176, pp. 861–77, 2015.

5   Böhme, M. *et al*, 'The Reconstruction of Early and Middle Miocene Climate and Vegetation in Southern Germany as Determined from the Fossil Wood Flora', *Palaeogeography, Palaeoclimatology, Palaeoecology*, Vol. 253, pp. 91–114, 2007.

6   Henry, A. and McIntyre, M., 'The Swamp Cypresses, *Glyptostrobus* of China and Taxodium of America, with Notes on Allied Genera', *Proceedings of the Royal Irish Academy*, Vol. 37, pp. 90–116, 1926.

7   Meller, B. *et al*, 'Middle Miocene Macro Floral Elements from the

Lavanttal Basin, Austria, Part 1, *Ginkgo adiantoides* (Unger) Heer', *Austrian Journal of Earth Sciences*, Vol. 108, pp. 185–98, 2015.

CHAPTER 16

1   Antoine, P. O. and Becker, D., 'A Brief Review of Agenian Rhinocerotids in Western Europe', *Swiss Journal of Geoscience*, Vol. 106, pp. 135–46, 2013.

2   Hooker, J. J. and Dashzeveg, D., 'The Origin of Chalicotheres (Perrisodactyla, Mammalia)', *Palaeontology*, Vol. 47, pp. 1363–68, 2004.

3   Sembrebon, G. *et al*, 'Potential Bark and Fruit Browsing as Revealed by Mibrowear Analysis of the Peculiar Clawed Herbivores Known as Chalicotheres (Perrisodactyla, Chalioctheroidea)', *Journal of Mammalian Evolution*, Vol. 18, pp. 33–55, 2010.

4   Barry, J. C. *et al*, 'Oligocene and Early Miocene Ruminants (Mammalia:Artiodactyla) from Pakistan and Uganda', *Palaeontologia Electronica*, Vol. 8, 2005.

5   Mitchell, G. and Skinner, J. D., 'On the Origin, Evolution and Phylogeny of Giraffes *Giraffa camelopardalis*', *Transactions of the Royal Society of South Africa*, Vol. 58, pp. 51–73, 2010.

6   Fossilworks: *Eotragus*.

7   Van der Made, J. and Mazo, A. V., 'Proboscidean Dispersal from Africa towards Western Europe', in Reumer, J. W. F. *et al* (eds.), 'Advances in Mammoth Research', *Proceedings of the Second International Mammoth Conference*, Rotterdam, 16–20 May 1999, 2003.

8   Wang, L.-H. and Zhang, Z.-Q., 'Late Miocene *Cervavitus noborossiae* (Cervidae, Artiodactyla) from Lantian, Shaanxi Province', *Vetebrata PalAsiatica*, Vol. 52, pp. 303–15, 2013.

9   Menecart, B., 'The Ruminantia (Mammalia, Certiodactyla) of the Oligocene to the Early Miocene of Western Europe: Systematics, Palaeoecology and Palaeobiogeography', PhD thesis 1756, University of Fribourg, 2012.

10   Garćes, M. *et al*, 'Old World First Appearance Datum of "Hipparion" Horses: Late Miocene Large Mammal Dispersal and Global Events', *Geology*, Vol. 25, pp. 19–22, 1997.

11   Agusti, J., 'The Biotic Environments of the Late Miocene Hominids', in Henke and Tattersal (eds), *Handbook of Palaeoanthropology*, Vol. 1, Ch. 5, Springer Reference, 2007.

12   Johnson, W. E. *et al*, 'The Late Miocene Radiation of Modern Felidae: A Genetic Assessment', *Science*, Vol. 311, pp. 73–77, 2006.

13   López-Antoňanzas, R. *et al*, 'New Species of Hispanomys (Rodentia, Cricetodontinae) from the Upper Miocene of Ballatones (Madrid, Spain)', *Zoological Journal of the Linnean Society*, Vol. 160, pp. 725–27, 2010.

14   Salesa, M. J. *et al*, 'Inferred Behaviour and Ecology of the Primitive Sabre-Toothed Cat *Paramachairodus ogygia* (Felidae, Machairodontinae) from the Late Miocene of Spain', *Journal of Zoology*, Vol. 268, pp. 243–54, 2006. Salesa, M. J. *et al*, 'First Known Complete Skulls of the Scimitar-Toothed Cat *Machairodus aphanistus* (Felidae, Carnivora) from the Spanish Late Miocene Site of Batallones–1', *Journal of Vertebrate Palaeontology*, Vol. 24, No. 4, pp. 957–69, 2004.

15   Sotnikova, M. and Rook, L., 'Dispersal of the Canini (Mammalia, Canidae: Caninae) across Eurasia during the Late Miocene to Early Pleistocene', *Quaternary International*, Vol. 212, pp. 86–97, 2010.

16   AFP, 'First Python Fossil Unearthed in Germany', 17 October 2011.

17   Mennecart, B. *et al*, 'A New Late Agenian (MN2a, Early Miocene) Fossil Assemblage from Wallenreid, (Molasse Basin, Canton Fribourg, Switzerland)', *Palaeontologische Zeitschrift*, Vol. 90, pp. 101–23, 2015. Kuch, U. *et al*, 'Snake Fangs from the Lower Miocene of Germany: Evolutionary Stability of Perfect Weapons', *Naturwissenschaften*, Vol. 93, pp. 84–87, 2006.

18   Evans, S. E. and Klembara, J., 'A Choristeran Reptile (reptilian:Diapsida) from the Lower Miocene of Northwest Bohemia (Czech Republic)', *Journal of Vertebrate Palaeontology*, Vol. 25, pp. 171–84, 2005.

CHAPTER 17

1   Darwin, C., *The Descent of Man, and Selection in Relation to Sex*, John Murray, London, 1871.

2   Begun, D., *The Real Planet of the Apes: A New Story of Human Origins*. Princeton University Press, Princeton, 2015.

3   *Ibid*.

4   *Ibid*.

5   Stevens, N. J., 'Palaeontological Evidence for an Oligocene Divergence between Old World Monkeys and Apes', *Nature*, Vol. 497, pp. 611–14, 2013.

6   Begun, D., *The Real Planet of the Apes: A New Story of Human Origins*, Princeton University Press, Princeton, 2015.

7   *Ibid*.

CHAPTER 18

1   *Ibid*.

2   *Ibid*.

3   Bernor, R. L., 'Recent Advances on Multidisciplinary Research at Rudabábanya, Late Miocene (MN9), Hungary', *Palaeontolographica Italica*, Vol. 89, pp. 3–36, 2002.

4   Begun, D., *The Real Planet of the Apes: A New Story of Human Origins*. Princeton University Press, Princeton, 2015.

5   *Ibid*.

6   Fuss, J. *et al*, 'Potential Hominin Affinities of *Graecopithecus* from the Late Miocene of Europe', *PLOS ONE*, Vol. 12, No. 5, 2017.

7   Böhme, M. *et al*, 'Messinian Age and Savannah Environment of the Possible Hominin *Graecopithecus* from Europe', *PLOS ONE*, Vol. 12, No. 5, 2017.

8   Gierliński, G. D., 'Possible Hominin Footprints from the Late Miocene (c. 5.7 Ma) of Crete?', *Proceedings of the Geologist's Association*, Vol. 128, Issues 5–6, pp. 697–710, 2017.

CHAPTER 19

1   Reyjol, Y. *et al*, 'Patterns in Species Richness and Endemism of European Freshwater Fish', *Global Ecology and Biogeography*, 15 December 2006.

2   Frimodt, C., *Multilingual Illustrated Guide to the World's Commercial Coldwater Fish*, Fishing News Books, Osney Mead, Oxford, 1995.

3   Venczel, M. and Sanchiz, B., 'A Fossil Plethodontid Salamander from the Middle Miocene of Slovakia (Caudata, Plethodontidae)', *Amphibia-Reptilia*, Vol. 26, pp. 408–11, 2005.

4   Naish, D., 'The Korean Cave Salamander', *Scientific American* blog, 18 August 2015.

CHAPTER 20

1   Stroganov, A. N., 'Genus *Gadus* (Gadidae): Composition, Distribution, and Evolution of Forms', *Journal of Ichthyology*, Vol. 55, pp. 319–36, 2015.

CHAPTER 21

1   Willis, K. J. and McElwain, J. C., *The Evolution of Plants*, (2nd ed.), Oxford University Press, Oxford, 2014.
2   Cadbury, D., *Terrible Lizard: The First Dinosaur Hunters and the Birth of a New Science,* Henry Holt, New York, 2000.
3   Owen, R., 'On the Fossil Vertebrae of a Serpent (*Laophis crotaloïdes*, Ow.) Discovered by Capt. Spratt, R. N., in a Tertiary Formation at Salonica', *Quarterly Journal of the Geological Society*, Vol. 13, pp. 197–98, 1857.
4   *Ibid*.
5   Boev, Z. and Koufous, G., 'Presence of *Pavo bravardi* (Gervais, 1849) (Aves, Phasianidae) in the Ruscinian Locality of Megalo Emvolon, Macedonia, Greece', *Geologica Balcanica*, Vol. 30, pp. 60–74, 2000.
     Pappas, S., 'Biggest Venomous Snake Ever Revealed in New Fossils', *Live Science*, 6 November 2014.
6   Georgalis, G. *et al*, 'Rediscovery of *Laophis crotaloides*—The World's Largest Viper', *Journal of Vertebrate Palaeontology Programme and Abstracts Book*, Berlin, 2014.
7   Pérez-García, A. *et al*, 'The Last Giant Continental Tortoise of Europe: A Survivor in the Spanish Pleistocene Site of Fonelas P-1', *Palaeogeography, Palaeoclimatology, Palaeoecology*, Vol. 470, pp. 30–39, 2017.
8   Bibi, F. *et al*, 'The Fossil Record and Evolution of Bovidae: State of the Field', *Palaeontologia Electronica*, No. 12(3) 10A, 2009.
9   Pimiento, C. and Balk, M. A., 'Body-Size Trends of the Extinct Giant Shark *Carcharocles megalodon*: A Deep-Time Perspective on Marine Apex Predators', *Paleobiology*, Vol. 41, No. 3, pp. 479–90, 2015.
10   Larramendi, A., 'Shoulder Height, Body Mass and Shape of Proboscideans', *Acta Palaeontologica Polonica*, Vol. 61, No. 3, pp. 537–74, 2016.
11   Van der Made, J. and Mazo, A. V., 'Proboscidean Dispersal from Africa towards Western Europe', in Reumer, J. W. F. *et al* (eds.), 'Advances in Mammoth Research', *Proceedings of the Second International Mammoth Conference*, Rotterdam, 16–20 May 1999.

12 Azzaroli, A., 'Quaternary Mammals and the "End-Villafranchian" Dispersal Event—A Turning Point in the History of Eurasia', *Palaeogeography, Palaeoclimatology, Palaeoecology*, Vol. 44, pp. 117–39, 1983.

13 Sotnikova, M. and Rook, L., 'Dispersal of the Canini (Mammalia, Canidae, Caninae) across Eurasia during the Late Miocene to Early Pleistocene', *Quaternary International*, Vol. 212, pp. 86–97, 2010.

CHAPTER 22

1 Lisiecki, L. E. and Raymo, M. E., 'A Pliocene-Pleistocene Stack of 57 Globally Distributed Benthic $\delta^{18}O$ Records', *Paleoceanography and Paleoclimatology*, 18 January 2005.

2 Blondel, J. *et al*, *The Mediterranean Region: Biological Diversity in Space and Time*, Oxford University Press, Oxford, 2010.

3 *Ibid*.

4 Rook, L. and Martinez-Navarro, B., 'Villafranchian: The Long Story of a Plio-Pleistocene European Large Mammal Biochronologic Unit', *Quaternary International*, Vol. 219, pp. 134–44, 2010.

5 Arribas, A. *et al*, 'A Mammalian Lost World in Southwest Europe during the Late Pliocene', *PLOS ONE*, Vol. 4, No. 9, 2009.

6 Turner, A. *et al*, 'The Giant Hyena, *Pachycrocuta brevirostris* (Mammalia, Carnivora, Hyaenidae), *Geobios*, Vol. 29, pp. 455–86, 1995.

7 Croitor, R., 'Early Pleistocene Small-Sized Deer of Europe', *Hellenic Journal of Geosciences*, Vol. 41, pp. 89–117, 2006.

8 Rook, L. and Martinez-Navarro, B., 'Villafranchian: The Long Story of a Plio-Pleistocene European Large Mammal Biochronologic Unit', *Quaternary International*, Vol. 219, pp. 134–44, 2010.

9 *Ibid*.

CHAPTER 23

1 Fisher, R. A., *The Genetical Theory of Natural Selection*, Clarendon Press, Oxford, 1930.

2 Gray, A., 'Mammalian Hybrids', Commonwealth Agriculture Bureaux, Edinburgh, Technical Publication No. 10, 1972.

3 Mallet, J., 'Hybridisation as an Invasion of the Genome', *Trends in Ecology and Evolution*, Vol. 20, pp. 229–37, 2005.

4   Kumar, V. *et al*, 'The Evolutionary History of Bears Is Characterised by Gene Flow across Species', *Scientific Reports* 7, Article No. 46487, 2017.
5   Palkopoulou, E. *et al*, 'A Comprehensive Genomic History of Extinct and Living Elephants', PubMed, National Institute of Health, 13 March 2018.
6   López Bosch, D., 'Hybrids and Sperm Thieves: Amphibian Kleptons', *All You Need Is Biology*, blog, 24 July 2016.
7   Gautier, M. *et al*, 'Deciphering the Wisent Demographic and Adaptive Histories from Individual Whole-Genome Sequences', *Biological Journal of the Linnean Society. Mol. Biol. Evol.*, Vol. 33, No. 11, pp. 2801–14, 2016.
8   Mallet, J., 'Hybridisation as an Invasion of the Genome', *Trends in Ecology and Evolution*, Vol. 20, pp. 229–37, 2005.
9   'Funny Creature "Toast of Botswana"', BBC News, 3 July 2000.
10 Darwin, C., *What Mr. Darwin Saw in His Voyage Round the World in the Ship 'Beagle'*, Harper & Bros., New York, 1879.
11 Hermansen, J. S. *et al*, 'Hybrid Speciation in Sparrows 1: Phenotypic Intermediacy, Phenotypic Admixture and Barriers to Gene Flow', *Molecular Ecology*, Vol. 2, pp. 3812–22, 2011.
12 Vallego-Marin, M., 'Hybrid Species Are on the March—with the Help of Humans', *The Conversation*, 31 May 2016.
    Noble, L., 'Hybrid "Super-Slugs" Are Invading British Gardens, and We Can't Stop Them', *The Conversation*, 19 April 2017.

CHAPTER 24

1   Sotnikova, M. and Rook, L., 'Dispersal of the Canini (Mammalia, Canidae: Caninae) across Eurasia during the Late Miocene to Early Pleistocene', *Quaternary International*, Vol. 212, pp. 86–97, 2010.
2   Ferring, R. *et al*, 'Earliest Human Occupations at Dmanisi (Georgian Caucasus) Dated to 1.85–1.78 Ma.', *PNAS*, Vol. 108, pp. 10432–36, 2013.
3   Lordkipanidze, D. *et al*, 'Postcranial Evidence from Early Homo from Dmanisi, Georgia', *Nature*, Vol. 449, pp. 305–10, 2007.
4   Lordkipanidze, D. *et al*, 'The Earliest Toothless Hominin Skull', *Nature*, Vol. 434, pp. 717–18, 2005.
5   Bower, B., 'Evolutionary Back Story: Thoroughly Modern Spine Supported Human Ancestor', *Science News*, Vol. 169. p. 275, 2009.
6   Mourer-Chauviré, C, and Geraads, D., 'The Struthionidae and Pelagornithidae (Aves: Struthioniformes, Odontopterygiformes) from

the Late Pliocene of Ahl Al Oughlam, Morocco', *Semantic Scholar*, 2008.

7   Fernández-Jalvo, Y. *et al*, 'Human Cannibalism in the Early Pleistocene of Europe (Gran Dolina, Sierra de Atapuerca, Burgos, Spain)', *Journal of Human Evolution*, Vol. 37, pp. 591–622, 1999.

8   Ashton, N. *et al*, 'Hominin Footprints from Early Pleistocene Deposits at Happisburgh, UK', *PLOS ONE*, 7 February 2014.

9   Wutkke, M., 'Generic Diversity and Distributional Dynamics of the Palaeobatrachidae (Amphibia: Anura)', *Palaeodiversity and Palaeoenvrinonments*, Vol. 92, No. 3, pp. 367–95, 2012.

CHAPTER 25

1   Golek, M. and Rieder, H., 'Erprobung der Altpalaolithischen Wurfspeere vol Schöningen', *Internationale Zeitschrift für Geschichte des Sports*, 25, Academic Verlag Sankt Augustin, 1–12, 1999.

2   Kozowyk, P. *et al*, 'Experimental Methods for the Palaeolithic Dry Distillation of Birch Bark: Implications for the Origin and Development of Neandertal Adhesive Technology', *Scientific Reports*, Vol. 7, p. 8033, 2017.

3   Mazza, P. *et al*, 'A New Palaeolithic Discovery: Tar-Hafted Stone Tools in a European Mid-Pleistocene Bone-Bearing Bed', *Journal of Archaeological Science*, Vol. 33, pp. 1310–18, 2006.

4   'The First Europeans—One Million Years Ago', *BBC Science and Nature*.

5   King, W., 'The Reputed Fossil Man of the Neanderthal', *Quarterly Journal of Science*, Vol. 1, p. 96, 1864.

6   Froehle, A. W. and Churchill, S. E., 'Energetic Competition between Neandertals and Anatomically Modern Humans', *PaleoAnthropology*, pp. 96–116, 2009.
Papagianni, D. and Morse, M., *The Neanderthals Rediscovered: How Modern Science Is Rewriting Their Story*, Thames & Hudson, London, 2013.
Bocherens, H., 'Isotopic Evidence for Diet and Subsistence Pattern of the Saint-Césaire I Neanderthal: Review and Use of a Multi-Source Mixing Model', *Journal of Human Evolution*, Vol. 49, No. 1, pp. 71–87, 2005.

7   Hoffecker, J. F. 'The Spread of Modern Humans in Europe', PNAS, Vol. 106, pp. 16040–45, 2009.

8   Boquet-Appel, J. P. and Degioanni, A., 'Neanderthal Demographic Estimates', *Current Anthropology*, Vol. 54, Issue 8, pp. 202–13, 2013.

9   Bergström, A. and Tyler-Smith, C., 'Palaeolithic Networking', *Science*, Vol. 358 (6363), pp. 586–87, 2017.

10   Tattersall, I., *The Strange Case of the Rickety Cossack and other Cautionary Tales from Human Evolution*, Palgrave Macmillan, New York, 2015.

11   Laleuza-Fox, C. *et al*, 'A Melanocortin 1 Receptor Allele Suggests Varying Pigmentation Among Neanderthals', *Science*, Vol. 318 (5855), pp. 1453–55, 2007.

12   Pierce, E. *et al*, 'New Insights into Differences in Brain Organization between Neanderthals and Anatomically Modern Humans', *Proceedings of the Royal Society (B)*, 280: 20130168, 2013.

13   Schwartz, S., 'The Mourning Dawn: Neanderthal Funerary Practices and Complex Response to Death', *HARTS and Minds*, Vol. 1, No. 3, 2013–14.

14   Hoffman, D. L. *et al*, 'U-Th Dating of Carbonate Crusts Reveals Neandertal Origin of Iberian Cave Art', *Science*, Vol. 359, pp. 912–15, 2018.

15   Radovčić, D., 'Evidence for Neandertal Jewelry: Modified White-Tailed Eagle Claws at Krapina', *PLOS ONE*, 11 March 2015.

16   Joubert, J. *et al*, 'Early Neanderthal Constructions Deep in Bruniquel Cave in Southwestern France', *Nature*, Vol. 534, pp. 111–14, 2016.

17   Lascu, C., *Piatra Altarului*, no publisher, undated.

18   Engelhard, M., *Ice Bear: The Cultural History of an Arctic Icon*, University of Washington Press, Washington, 2016.

19   Hingham, T. *et al*, 'The Timing and Spatiotemporal Patterning of Neanderthal Disappearance', *Nature*, Vol. 512, pp. 306–09, 2014.

CHAPTER 26

1   Hershkovitz, I., *et al*, 'The Earliest Modern Humans Outside Africa', *Science*, Vol. 359, pp. 456–59, 2018.
     Richter, D. *et al*, 'The Age of the Hominin Fossils from Jebel Irhoud, Morocco, and the Origins of the Middle Stone Age', *Nature*, Vol. 546, pp. 293–96, 2017.
     Fu, Q. *et al*, 'Genome Sequence of a 45,000-Year-Old Modern Human from Western Siberia', *Nature*, Vol. 514, pp. 445–49, 2016.

2   Fu, Q. *et al*, 'The Genetic History of Ice-age Europe', *Nature*, Vol. 534, pp. 200–05, 2016.

3   Fu, Q. *et al*, 'An Early Modern Human Ancestor from Romania with a Recent Neanderthal Ancestor', *Nature*, Vol. 524, pp. 216–19, 2015.

4   *Ibid*.
5   Hartwell Jones, G., *The Dawn of European Civilisation*, Gilbert and Rivington, London, 1903.
6   Green, R. E. *et al*, 'Draft Full Sequence of Neanderthal Genome', *Science*, Vol. 328, pp. 710–22, 2010.
7   Mendez, F. L. *et al*, 'The Divergence of Neandertal and Modern Human Y Chromosomes', *American Journal of Human Genetics*, Vol. 98, No. 4, pp. 728–34, 2016.
8   Sankararaman, S., *et al*, 'The Genomic Landscape of Neanderthal Ancestry in Present-day Humans', *Nature*, Vol. 507, pp. 354–57, 2014.
9   Bennazi, S. *et al*, 'Early Dispersal of Modern Humans in Europe and Implications for Neanderthal Behaviour', *Nature*, Vol. 279, pp. 525–28, 2011. Hingham, T. *et al*, 'The Earliest Evidence of Anatomically Modern Humans in Northwestern Europe', *Nature*, Vol. 479, pp. 521–24, 2011.
10   Vernot, B. and Akey, J. M., 'Resurrecting Surviving Neandertal Lineages from Modern Human Genomes', *Science*, Vol. 343, pp. 1017–21, 2014.
11   Fu, Q. *et al*, 'The Genetic History of Ice-age Europe', *Nature*, Vol. 534, pp. 200–05, 2016.
12   Yong, E., 'Surprise! 20 Percent of Neanderthal Genome Lives on in Modern Humans, Scientists Find', *National Geographic*, 29 January 2014.

CHAPTER 27

1   Dvorsky, G, 'A 40,000 Year-Old Sculpture Made Entirely from Mammoth Ivory', *Gizmodo*, 2 August, 2013.
2   Quiles, A. *et al*, 'A High-Precision Chronological Model for the Decorated Upper Palaeolithic Cave of Chauvet-Pont d'Arc, Ardéche, France', *PNAS*, Vol. 113, pp. 4670–75, 2016.
3   Thalmann, O. *et al*, 'Complete Mitochondrial Genomes of Ancient Canids Suggest a European Origin of Domestic Dogs', *Science*, Vol. 342, Issue 6160, pp. 871–74, 2013.
4   Sotnikova, M. and Rook, L., 'Dispersal of the Canini (Mammalia, Canidae, Caninae) across Eurasia during the Late Miocene to Early Pleistocene', *Quaternary International*, Vol. 212, pp. 86–97, 2010.
5   Dugatkin, L. A. and Trutt, L., *How to Tame a Fox*, University of Chicago Press, Chicago, 2017.
6   Napierala, H., and Uerpmann, H-P., 'A "New" Palaeolithic Dog from

Central Europe', *International Journal of Osteoarchaeology*, Vol. 22, pp. 127–37, 2010.

7   Frantz, L. A. F., *et al*, 'Genomic and Archaeological Evidence Suggest a Dual Origin of Domestic Dogs', *Science*, Vol. 352, Issue 6290, pp. 1228–31, 2016.
     Botigué, L. R., *et al,* 'Ancient European Dog Genomes Reveal Continuity Since the Early Neolithic', *Nature Communications*, Vol. 8, Article No. 16082, 2017.

CHAPTER 28

1   Callaway, E., 'Elephant History Rewritten by Ancient Genomes', *Nature*, News, 16 September 2016.

2   Palkopoulou, E. *et al*, 'A Comprehensive Genomic History of Extinct and Living Elephants', *PNAS*, 26 February 2018.

3   Thieme, H. and Veil, S., 'Neue Untersuchungen zum eemzeitlichen Elefanten-Jagdplatz Lengingen', Ldkg. Verden. *Die Kunde*, Vol. 236, pp. 11–58, 1985.

4   Geer, A. van der, *et al*, *Evolution of Island Mammals*, Wiley Blackwell, UK, 2010.

CHAPTER 29

1   Pushkina, D., 'The Pleistocene Easternmost Distribution in Eurasia of the Species Associated with the Eemian *Palaeloxodon antiquus* Assemblage', *Mammal Reviews*, Vol. 37, pp. 224–45, 2007.

2   Pulcher, E., 'Erstnachweis des europaischen Wilkdesels (*Equus hydruntius*, Regalia, 1907) im Holozan Österreichs', 1991.

3   Naito, Y. I. *et al*, 'Evidence for Herbivorous Cave Bears (*Ursus spelaeus*) in Goyet Cave, Belgium: Implications for Palaeodietary Reconstruction of Fossil Bears Using Amino Acid δ^{15}N Approaches', *Journal of Quaternary Science*, Vol. 31, pp. 598–606, 2016.

4   Pacher, M. and Stuart, A., 'Extinction Chronology and Palaeobiology of the Cave Bear (*Ursus spelaeus*)', *Boreas*, Vol. 35, Issue 2, pp. 189–206, 2008.

5   MüS, C. and Conard, N. J., 'Cave Bear Hunting in the Hohle Fels, a Cave Site in the Ach Valley, Swabian Jura', *Revue de Paléobiologie*, Vol. 23, Issue 2, pp. 877–85, 2004.

6   Gonzales, S. *et al*, 'Survival of the Irish Elk into the Holocene', *Nature*, Vol. 405, pp. 753–54, 2000.

7   Kirillova, I. V., 'On the Discovery of a Cave Lion from the Malyi Anyui River (Chukotka, Russia)', *Quaternary Science Reviews*, Vol. 117, pp. 135–51, 2015.

8   Bocherens, H. *et al*, 'Isotopic Evidence for Dietary Ecology of Cave Lion (*Panthera spelaea*) in North-Western Europe: Prey Choice, Competition and Implications for Extinction', *Quaternary International*, Vol. 245, pp. 249–61, 2011.

9   Cuerto, M. *et al*, 'Under the Skin of a Lion: Unique Evidence of Upper Palaeolithic Exploitation and Use of Cave Lion (*Panthera spelaea*) from the Lower Gallery of La Garma (Spain)', *PLOS ONE*, Vol. 11, Issue 10, Article no. e0163591, 2016.

10   Rohland, N. *et al*, 'The Population History of Extant and Extinct Hyenas', *Molecular Biology and Evolution*, Vol. 22, pp. 2435–43, 2005.

11   Varela, S. *et al*, 'Were the Late European Climatic Changes Responsible for the Disappearance of the European Spotted Hyena Populations? Hindcasting a Species Geographic Distribution across Time', *Quaternary Science Reviews*, Vol. 29, pp. 2027–35, 2010.

12   Diedrich, C. G., 'Late Pleistocene Leopards across Europe—Northernmost European German Population, Highest Elevated Records in the Swiss Alps, Complete Skeletons in the Bosnia Herzegovina Dinarids and Comparison to the Ice-Age Cave Art', *Quaternary Science Reviews*, Vol. 76, pp. 167–93, 2013.
     Sommer, R. S. and Benecke, N., 'Late Pleistocene and Holocene Development of the Felid Fauna (Felidae) of Europe: A Review', *Journal of Zoology*, Vol. 269, pp. 7–19, 2005.

CHAPTER 30

1   Gupta, S. *et al*, 'Two-Stage Opening of the Dover Strait and the Origin of Island Britain', *Nature Communications*, Vol. 8, Article No. 15101, 2017.

2   Kahlke, R. D., 'The Origin of Eurasian Mammoth Faunas (*Mammuthus, Coelodonta* Faunal Complex)', *Quaternary Science Reviews*, Vol. 96, pp. 32–49, 2012.

3   Todd, N. E., 'Trends in Proboscidean Diversity in the African Cenozoic', *Journal of Mammalian Evolution*, Vol. 13, pp. 1–10, 2006.

4   Stuart, A. J. *et al*, 'The Latest Woolly Mammoths (*Mammuthus primi-genius* Blumenbach) in Europe and Asia: A Review of the Current Evidence', *Quaternary Science Reviews*, Vol. 21, pp. 1559–69, 2002.

5   Palkopoulou, E. *et al*, 'Holarctic Genetic Structure and Range Dynamics in the Woolly Mammoth', *Proceedings of the Royal Society B*, Vol. 280, Issue 1770, 2013.
    Lister, A. M., 'Late-Glacial Mammoth Skeletons (*Mammuthus primigenius*) from Condover (Shropshire, UK): Anatomy, Pathology, Taphonomy and Chronological Significance', *Geological Journal*, Vol. 44, pp. 447–79, 2009.

6   Stuart, A. J. *et al*, 'The Latest Woolly Mammoths (*Mammuthus primi-genius* Blumenbach) in Europe and Asia: A Review of the Current Evidence', *Quaternary Science Reviews*, Vol. 21, pp. 1559–69, 2002.

7   Boeskorov, G. G., 'Some Specific Morphological and Ecological Features of the Fossil Woolly Rhinoceros (*Coelodonta antiquitatis* Blumenbach 1799)', *Biology Bulletin*, Vol. 39, Issue 8, pp. 692–707, 2012.

8   Jacobi, R. M. *et al*, 'Revised Radiocarbon Ages on Woolly Rhinoceros (*Coelodonta antiquitatis*) from Western Central Scotland: Significance for Timing the Extinction of Woolly Rhinoceros in Britain and the Onset of the LGM in Central Scotland', *Quaternary Science Reviews*, Vol. 28, pp. 2551–56, 2009.

9   Shpansky, A. V. *et al*, 'The Quaternary Mammals from Kozhamzhar Locality, (Pavlodar Region, Kazakhstan)', *American Journal of Applied Science*, Vol. 13, pp. 189–99, 2016.

10 Reumer, J. W. F. *et al*, 'Late Pleistocene Survival of the Saber-Toothed Cat *Homotherium* in Northwestern Europe', *Journal of Vertebrate Paleontology*, Vol. 23, pp. 260–62, 2003.

11 A fuller discussion of the decline of the sabre-tooths can be found in: Macdonald, D. and Loveridge, A., *The Biology and Conservation of Wild Felids*, Oxford University Press, Oxford, 2010.

CHAPTER 31

1   Guthrie, R. D., *The Nature of Paleolithic Art*, University of Chicago Press, Chicago, 2005.

2   Quiles, A, *et al*, 'A High-Precision Chronological Model for the Decorated Upper Palaeolithic Cave of Chauvet-Pont d'Arc, Ardéche, France', *PNAS*, Vol. 113, pp. 4670–75, 2016.

3 Guthrie, R. D., *The Nature of Paleolithic Art*, University of Chicago Press, Chicago, 2005, pp. 276–96.
4 *Ibid*, p. 324.
5 Schmidt, I., *Solutrean Points of the Iberian Peninsula: Tool Making and Using Behaviour of Hunter-Gatherers during the Last Glacial Maximum*, British Archaeological Reports, Oxford, 2015.

CHAPTER 32

1 Tallavaara, M. L. *et al*, 'Human Population Dynamics in Europe over the Last Glacial Maximum', *PNAS*, Vol. 112, Issue 27, pp. 8232–37, 2015.
2 Sommer, R. S. and Benecke, N., 'Late Pleistocene and Holocene Development of the Felid Fauna (Felidae) of Europe: A Review', *Journal of Zoology*, Vol. 269, Issue 1, pp. 7–19, 2006.
3 Heptner, V. G. and Sludskii, A. A., *Mammals of the Soviet Union, Vol. II*, Part 2, 'Carnivora (Hyaenas and Cats)', Leiden, New York, 1992.
   Üstay, A. H., *Hunting in Turkey*, BBA, Istanbul, 1990.
4 Rohland, N. *et al*, 'The Population History of Extant and Extinct Hyenas', *Molecular Biology and Evolution*, Vol. 22, Issue 12, pp. 2435–43, 2005.
5 Fu, Q. *et al*, 'The Genetic History of Ice Age Europe', *Nature*, Vol. 534, pp. 200–05, 2016.
6 Schmidt, K., 'Göbekli Tepe—Eine Beschreibung der wichtigsten Befunde erstellt nach den Arbeiten der Grabungsteams der Jahre 1995– 2007', in *Erste Tempel—Frühe Siedlungen, 12000 Jahre Kunst und Kultur*, Oldenburg, 2009.

CHAPTER 33

1 Huntley, B., 'European Post-Glacial Forests: Compositional Changes in Response to Climatic Change', *Journal of Vegetation Science*, Vol. 1, pp. 507–18, 1990.
2 Zeder, M. A., 'Domestication and Early Agriculture in the Mediterranean Basin: Origins, Diffusion, and Impact, *PNAS*, Vol. 105, Issue 33, pp. 11597–604, 2008.
3 Fagan, B., *The Long Summer: How Climate Changed Civilisation*, Granta Books, London, 2004.
4 Zilhao, J., 'Radiocarbon Evidence for Maritime Pioneer Colonisation at

the Origins of Farming in West Mediterranean Europe', *PNAS*, Vol. 98, pp. 14180–85, 2001.

5   Frantz, A .C., 'Genetic Evidence for Introgression Between Domestic Pigs and Wild Boars (*Sus scrofa*) in Belgium and Luxembourg: A Comparative Approach with Multiple Marker Systems', *Biological Journal of the Linnean Society*, Vol. 110, pp. 104–15, 2013.

6   Park, S. D. E. *et al*, 'Genome Sequencing of the Extinct Eurasian Wild Aurochs, *Bos primigenius*, Illuminates the Phylogeography and Evolution of Cattle, *Genome Biology*, Vol. 16, p. 234, 2015.

CHAPTER 34

1   Bramanti, B. *et al*, 'Genetic Discontinuity Between Local Hunter-Gatherers and Central Europe's First Farmers, *Science*, Vol. 326, pp. 137–40, 2009.

2   Downey, S. E. *et al*, 'The Neolithic Demographic Transition in Europe: Correlation with Juvenile Index Supports Interpretation of the Summed Calibrated Radiocarbon Date Probability Distribution (SCDPD) as a Valid Demographic Proxy', *PLOS ONE*, 9(8): e105730, 25 August 2014.

3   'Childe, Vere Gordon (1892–1957)', *Australian Dictionary of Biography*, Melbourne University Publishing, Melbourne, 1979.

4   Low, J., 'New Light on the Death of V. Gordon Childe', *Australian Society for the Study of Labour History*, undated, www.laborhistory.org.au/hummer/no-8/gordon-childe/

5   Green, K., 'V. Gordon Childe and the Vocabulary of Revolutionary Change', *Antiquity*, Vol. 73, pp. 97–107, 1961.

6   Stevenson, A., 'Yours (Unusually) Cheerfully, Gordon: Vere Gordon Childe's Letters to RBK Stevenson', *Antiquity*, Vol. 85, pp. 1454–62, 2011.

7   Editorial, *Antiquity*, Vol. 54, No. 210, p. 2, 1980.

8   Cieslak, M. *et al*, 'Origin and History of Mitochondrial DNA Lineages in Domestic Horses', *PLOS ONE*, 5(2): e15311, 2010.

9   *Ibid*.

10 Almathen, F. *et al*, 'Ancient and Modern DNA Reveal Dynamics of Domestication and Cross-Continental Dispersion of the Dromedary', *PNAS*, Vol. 113, pp. 6706–12, 2016.

11 Gunther, R. T., 'The Oyster Culture of the Ancient Romans', *Journal of the Marine Biological Association of the United Kingdom*, Vol. 4, pp. 360–65, 1897.

CHAPTER 35

1   Van der Geer, A. *et al*, *Evolution of Island Mammals: Adaptation and Extinction of Placental Mammals on Islands*, Wiley-Blackwell, New Jersey, 2010.

2   Lyras, G. A. *et al*, '*Cynotherium sardous*, an Insular Canid (Mammalia: Carnivora) from the Pleistocene of Sardinia (Italy), and its Origin', *Journal of Vertebrate Palaeontology*, Vol. 26, pp. 735–45, 2005.

3   Hautier, L. *et al*, 'Mandible Morphometrics, Dental Microwear Pattern, and Palaeobiology of the Extinct Belaric Dormouse *Hypnomys morpheus*', *Acta Palaeontologica Polonica*, Vol. 54, pp. 181–94, 2009.

4   Shindler, K., '*Discovering Dorothea: The Life of the Pioneering Fossil-Hunter Dorothea Bate*, Harper Collins, London, 2005.

5   Ramis, D. and Bover, P., 'A Review of the Evidence for Domestication of *Myotragus balearicus* Bate 1909 (Artiodactyla, Caprinae) in the Balearic Islands', *Journal of Archaeological Science*, Vol. 28, pp. 265–82, 2001.

CHAPTER 36

1   Hirst, J., *The Shortest History of Europe*, Black Inc, Melbourne, 2012.

2   Rokoscz, M., 'History of the Aurochs (*Bos Taurus primigenius*) in Poland', *Animal Genetic Resources Information*, Vol. 16, pp. 5–12, 1995.

3   *Ibid*.

4   Elsner, J. *et al*, 'Ancient mtDNA Diversity Reveals Specific Population Development of Wild Horses in Switzerland after the Last Glacial Maximum', *PLOS ONE*, 12(5): e0177458, 2017.

5   Sommer, R. S., 'Holocene Survival of the Wild Horse in Europe: A Matter of Open Landscape?' *Journal of Quaternary Science*, Vol. 26, Issue 8, pp. 805–12, 2011.

6   Van Vuure, C. T., 'On the Origin of the Polish Konik and Its Relation to Dutch Nature Management', *Lutra*, Vol. 57, pp. 111–30, 2014.

7   Gautier, M. *et al*, 'Deciphering the Wisent Demographic and Adaptive Histories from Individual Whole-Genome Sequences', *Biological Journal of the Linnean Society, Mol. Biol. Evol.*, Vol. 33, Issue 11, pp. 2801–14, 2016.

8   Vera, F. and Buissink, F., 'Wilderness in Europe: What Really Goes on between the Trees and the Beasts', Tirion Baarn (Netherlands), 2007.

9   Bashkirov, I. S., 'Caucasian European Bison', Moscow: Central Board for

Reserves, Forest Parks and Zoological Gardens, Council of the People's Commissars of the RSFSR, pp. 1–72, 1939. [In Russian.]

CHAPTER 37

1   Hoffman, G. S. *et al*, 'Population Dynamics of a Natural Red Deer Population over 200 Years Detected via Substantial Changes of Genetic Variation', *Ecololgy and Evolution*, Vol. 6, pp. 3146–53, 2016.
2   Fritts, S. H., *et al*, 'Wolves and Humans', in Mech, L. D. and Boitani, L. (eds), *Wolves: Behavior, Ecology and Conservation*, University of Chicago Press, Chicago, 2003.
3   Lagerås, C., *Environment, Society and the Black Death: An Interdisciplinary Approach to the Late Medieval Crisis in Sweden*, Oxbow Books, Oxford, 2016.
4   Albrecht, J. *et al,* 'Humans and Climate Change Drove the Holocene Decline of the Brown Bear', *Nature, Scientific Reports*, 7, Article No. 10399, 2017.
5   Engelhard, M., *Ice Bear: The Cultural History of an Arctic Icon*, University of Washington Press, Seattle, 2016
6   Zeder, M. A., 'Domestication and Early Agriculture in the Mediterranean Basin: Origins, Diffusion, and Impact', *PNAS*, Vol. 105, No. 33, pp. 11597–604, 2008.
7   Hard, J. J. *et al*, 'Genetic Implications of Reduced Survival of Male Red Deer *Cervus elaphus* under Harvest', *Wildlife Biology*, Vol. 2, Issue 4, pp. 427–41, 2006.

CHAPTER 38

1   Cunliffe, B., *By Steppe, Desert, and Ocean: The Birth of Eurasia*, Oxford University Press, Oxford, 2015.
2   Thompson, V. *et al*, 'Molecular Genetic Evidence for the Place of Origin of the Pacific Rat, *Rattus exulans*', *PLOS ONE*, 17 March 2014.

CHAPTER 39

1   Poole, K., *Extinctions and Invasions: A Social History of British Fauna*, chapter 18, 'Bird Introductions', Oxbow Books, Oxford, 2013.
2   *Ibid.*

3 'The History of the Pheasant', *The Field*, www.thefield.co.uk

4 Glueckstein, F., 'Curiosities: Churchill and the Barbary Macaques', *Finest Hour*, Vol. 161, 2014.

5 Masseti, M. *et al*, 'The Created Porcupine, *Hystrix cristata* L. 1758, in Italy', *Anthropozoologica*, Vol. 45, pp. 27–42, 2010.

6 Nykl, A. R., *Hispano-Arabic Poetry and Its Relations with the Old Provincal Troubadors*, John Hopkins University Press, Baltimore, 1946.

7 Fagan, B., *Fishing: How the Sea Fed Civilisation*, Yale University Press, New Haven, 2017.

CHAPTER 40

1 Montaigne, M., *Les Essais*, Abel the Angelier, Paris, 1598.

2 Pakenham, T., Reply in 'The Bastard Sycamore', *New York Review of Books* letters page, 19 January 2017.

3 Halamski, A. T., 'Latest Cretaceous Leaf Floras from Southern Poland and Western Ukraine', *Acta Palaeontologica*, Vol. 58, pp. 407–43, 2013.

4 Sheehy, E. and Lawton, C., 'Population Crash in an Invasive Species Following the Recovery of a Native Predator: The Case of the American Grey Squirrel and the European Pine Marten in Ireland', *Biodiversity and Conservation*, Vol. 23, Issue 3, pp. 753–74, 2014.

5 Bertolino, S. and Genovesi, P., 'Spread and Attempted Eradication of the Grey Squirrel (*Sciurus carolinensis*) in Italy, and Consequences for the Red Squirrel (*Sciurus vulgaris*) in Eurasia', *Biological Conservation*, Vol. 109, pp. 351–58, 2003.

6 Tizzani, P. *et al*, 'Invasive Species and Their Parasites: Eastern Cottontail Rabbit *Sylvilagus floridanus* and *Trichostrongylus affinis* (Graybill 1924) from Northwestern Italy', *Parasitological Research*, Vol. 113, pp. 1301–03, 2014.

7 Hohmann, U. *et al*, *Der Waschbär*, Oertel and Spörer, Reutlingen, 2001.

8 'Kangaroos run wild in France', *AFP*, 12 November 2003.

9 Mali, I. *et al*, 'Magnitude of the Freshwater Turtle Exports from the US: Long-Term Trends and Early Effects of Newly Implemented Harvesting Regimes', *PLOS ONE*, 9(1), E86478, 2014.

CHAPTER 41

1   Pierotti, R. and Fogg, B., *The First Domestication: How Wolves and Humans Co-evolved*, Yale University Press, New Haven, 2017.

2   Ó'Crohan, T., *The Islandman*, The Talbot Press, Dublin and Cork, 1929.

CHAPTER 42

1   Inger, R. *et al*, 'Common European Birds Are Declining Rapidly while Less Abundant Species' Numbers Are Rising', *Ecology Letters*, Vol. 18, pp. 28–36, 2014.

2   D W News, '"Dramatic" Decline in European Birds Linked to Industrial Agriculture', 4, May 2017.

3   Vogel, G., 'Where Have All the Insects Gone?', *Science*, 10 May 2017.

4   Ruiz, J., 'A New EU Agricultural Policy for People and Nature', *EUACTIV*, 28 April 2017.

5   *EIONET*, 'State of Nature in the EU: Reporting Under the Birds and Habitats Directives', 2015.

6   *Ibid.*

7   Tree, I., *Wilding: The Return of Nature to an English Farm,* Picador, London, 2018.

8   Herard, F. *et al*, '*Anoplophora glabripennis*—Eradication Programme in Italy', European and Mediterranean Plant Protection Organization, 2009.

9   Stafford, F., *The Long, Long Life of Trees*, Yale University Press, New Haven, 2016.

CHAPTER 43

1   Tauros Scientific Programme, taurosprogramme.com/tauros-scientific-programme/

2   Tacitus, C., *Germany and Its Tribes*, (translated by Church, A. J. and Brodribb, W. J.), Macmillan, London 1888.

3   Rice, P. H., 'A Relic of the Nazi Past Is Grazing at the National Zoo', United States Holocaust Memorial Museum, 3 April 2017.

4   Van Vuure, C. T., 'On the Origin of the Polish Konik and Its Relation to Dutch Nature Management', *Lutra*, Vol. 57, pp. 111–30, 2014.

5   *Ibid.*

CHAPTER 44

1 Choi, C., 'First Extinct Animal Clone Created', *National Geographic News*, 10 February 2009.
2 Revive & Restore, reviverestore.org
3 Pilcher, H., 'Reviving Woolly Mammoths Will Take More than Two Years', *BBC Earth*, 22 February 2017.
4 Rewilding Europe, rewildingeurope.com/background-and-goals/ urbanisation-and-land-abandonment/
5 *Ibid.*

ENVOI

1 Roemer, N., *German City, Jewish Memory: The Story of Worms*, UPNE, 2010.

Index